Prepare For The Military

SSGT NICHOLS

ISBN: 0983055653
ISBN-13: 978-0983055655

CONTENTS

FORWARD BY SSGT NICHOLS

I served in the United States Marine Corps and would not presume expertise in any other branch of the military. Yet even though my personal stories convey experiences from the Marine Corps, I have written this book with the hopes of providing guidance to anyone with a desire to serve, whether they choose the Air Force, Navy, Army, Coast Guard or Marine Corps. This book will help you understand how to get physically and mentally ready before you join the military, what to expect in Boot Camp, building a military career, how to become an officer, and dealing with the rigors of deployment (to name a few topics).

My name is SSGT Nichols, and the United States Marine Corps has defined a significant portion of my life.

Joining the Marine Corps was not my idea. My older brother Larry Nichols Jr. (Boe) first signed up in the Delayed Entry Program. Just prior to his departure for Boot Camp, Boe's recruiter talked me into joining the military via the Buddy Program. I had just graduated from Lutcher High School in Louisiana and truly had no idea what direction I was headed. I was working here and there trying to figure out exactly what to do with my life.

Growing up, no one ever really motivated me to think about my future. I considered the military, but at the time really did not want to leave my friends and family.

Joining the Marine Corps did not sound like a bad idea: steady income, world travel, San Diego beaches on the weekends, and a $1,000.00 signing bonus for joining through the Buddy Program. It was all good!

After several weeks of evaluations, testing, and medical exams, I was accepted for training. So off to Boot Camp I would go, not having any knowledge of recruit training nor having prepared at all.

Shortly into Boot Camp, around the third day, Boe suffered a major back injury and was discharged from the Marine Corps. I was given the chance to return home with him, but Boe reminded me that this opportunity, this challenge to become a Marine was once in a lifetime. I made the choice to change my life forever, not just for myself but for both of us. On May 5, 1995, I graduated from Boot Camp, earning one of the most coveted titles: a United States Marine.

In late 2000, I received orders to Marine Corps Recruit Depot (MCRD) San Diego to become a Marine Corps Drill Instructor. It was during my initial Senior Drill Instructor cycle that a film crew was embedded with my platoon. The filming of Platoon 1141 was featured in the subsequent release of *Black Friday: Dark Dawn.*

MCRD

In 2006, I deployed to Iraq in support of Operation Iraqi Freedom. Based in Al Asad Iraq, I completed 28 combat missions, 13 as a Platoon Sergeant and 15 as the Platoon Commander after my Lieutenant was severely injured in an IED attack.

On March 25, 2007 after 12 years of service, I was honorably discharged from the United States Marine Corps.

I wrote this book to share valuable information to anyone looking to join the military. While the purpose of this book is not to serve as an advocate for the military (the military is not a lifestyle for everyone), I would like you to consider the transformational aspects of joining.

By enlisting or accepting a commission in the military, you are volunteering to become a servant of this country. For me, serving this country is a very high calling. Serving this country is the overarching reason why anyone should join the military.

On a personal level beyond the desire to serve, consider some other factors. If you are a high school student or recent college graduate, your job options are going to be decidedly entry-level. There is no major corporation in the entire United States that will entrust you with the kind of responsibilities that the military expects you to handle from day one. As an enlistee, you will be handling equipment designed for combat. If you become a commissioned officer, you will be put in important management positions years before your civilian counterparts will see that level of responsibility.

Something else to consider. The kinds of jobs that are available to you in your teens and twenties are also the kinds of jobs that will be available to you in your fifties. That is not the case for the military. You have a very short window of opportunity, starting in your late teens and generally ending in your mid-thirties, when you can join the military. If you want to join the Marine Corps, your window of time is even shorter.

Ask a lot of veterans, and they will tell you that had they not been in the military, they would have certainly looked back at their lives with some regret. There are a lot of people in their forties and fifties doing the same things now as they were in their twenties. Many regret not doing something more exciting when they were young and had the chance to do so. Many things in life can wait, but not the military. Blow the window to join, fail to seize the opportunity, and you will never get another chance.

In the military, you will be surrounded by professionals who take their jobs very seriously. These men and women will be mentors and can help break you away from any bad influences of civilian life. For example, there is no tolerance for drug or alcohol abuse.

You will find that your own expectations for personal performance rise during your time in the military. This is part of being held to a higher standard every day by professionals dedicated to serve. You will begin to notice that the once acceptable standards of the civilian world are no longer tolerable.

The military is a dangerous job. You can get hurt or killed. You may deploy to combat zones and be overseas for months or years. There is no avoiding this fact. If you sign up for the military, you are taking a risk. But as with many endeavors worthwhile in life, risk is part of the equation. And with the military, the payoff is the chance to experience the adventure of a lifetime. You will be exposed to diverse cultures in far-away places and lifestyles that are completely different from main street USA. You will get to visit parts of the world that most people only see when they go to the movies.

The military provides you with a glorious opportunity to break with your past. Some parents continue to baby their children through adulthood. If you are in high school or college right now, just take a look around and see how many of your peers still behave with a childlike mindset.

Watch what happens the day you let your parents know that you have just signed up for the military. You will stop being a child very quickly. Indeed, after a few years of service, you will find that your parents, your teachers, in fact all grown-ups in your hometown and everywhere else, will treat you completely differently. The old saying that the military turns boys into men may be from a different era, but it still holds in spirit, as it takes a real man or woman to serve in the armed forces. The general population, including those who have never served themselves, instinctively know this to be true.

Joining the military is one way to get exposed to many different career paths and start earning some decent money while doing something exciting and adventurous. You will discover the limitless options you have in your life. Years later, you will look back at your time in the military with a sense of wonder and adventure that you may never experience again.

1 INTRODUCTION

Virtually everyone who has served in the military can relate to this story. It is the day before you are scheduled to board the plane for Boot Camp. Looking around, you see everything that you have known for your entire life. Yet you know this is the last day, ever, where everything is familiar. In your mind, you begin the withdrawal process, knowing that the next time you return here, nothing will be the same. The next day, as you head to the airport, the anxiety is palpable. Your heart is beating fast, your throat is dry. You are wondering if you made the right decision. Then, as you walk towards the plane, you take one last look back, and say goodbye to the world you have always known. Goodbye. Forever.

When I boarded that plane, I got to be with my brother (because I had enlisted in the Buddy Program). I distinctly remember the level of heightened anxiety in both my brother and I as we got closer to the reckoning. Emotions oscillated between happiness and mind numbing fear.

Once we landed in San Diego, I remember my heart sinking to the bottom of my chest. As we disembarked the plane, I told my brother that we should go to the bathroom (which later would be known as a "head") so we would not miss our chance to freshen up. We took our time, looking for one last moment at our undisciplined bodies in the mirror and saying to ourselves, here goes nothing.

We looked at each other, took a deep breath, and off we went. Ahead of us, walking down the airport terminal was the silhouette of a Marine Corps Drill Instructor. His raspy voice echoed and the words "Hurry the Freak Up!" came out of his mouth. We immediately went from a normal stroll in the park to a power walk (which is one step faster than a normal walk but one step slower than a run).

There seemed to be a lot of young men who were getting yelled at and being placed outside in a single file line, an arrangement colloquially known as "A to B". A to B is when you line up with your toes touching the heels of the recruit in front of you and you are literally breathing on his neck. It seemed like we were standing there forever. Once in a while, you would hear the Drill Instructor bark out commands from inside the terminal while he waited for others to arrive.

When all the recruits were present, we boarded a white government bus. Once seated, we were told to keep our heads down and were warned that we better not get caught looking around. It felt like we were on the way to a secret prison camp. I remember my brother and I sitting together and whispering to each other, wondering what was going on.

That seemed to be the fastest bus ride ever. You could just feel the tension in the air, because we all had an idea what was waiting for us once we got on the base. Soon the bus driver was talking with the guard at the gate of MCRD and we subsequently entered the base. Then it happened. The bus came to a screeching halt. Next thing we know, a Marine Corps Drill Instructor started moving us off the bus.

It was late at night, dark and cold, and we found ourselves standing on the iconic yellow footprints. Amidst the yelling and chaos, I remember asking myself what in the world was going on? Nothing made sense to me. Prior to this moment in my life, I had never really watched military movies, never had a family member in the military, and never participated in high school JROTC. Looking back now, I sometimes wonder if that was a mistake. As commands were barked out by hardened Marines, other recruits with training and experience under their belt seemed to be better prepared than me. They participated in JROTC, were raised in a military family or had the chance to get involved with the Marine Corps Delayed Entry Program. I did not have any prior experience but still was standing there in the mitts of swarming Drill Instructors. They were moving at the speed of light with voices that would give goose bumps to any sane person.

Raspy and mean looking, the last thing you wanted was their attention. To ensure I did not get any special attention, I followed the lead of others, which seemed to work well at the time. Have you ever heard of the saying "Fake it 'til you make it"? If not, now you have.

It did not take me long to catch on because I was willing to learn. I had always been a very motivated person, but had no structure and discipline growing up. The Marine Corps took charge of my life, and transformed it from the undisciplined world of freedom to the disciplined world of structure.

Thirteen weeks was going to be a lifetime. Why did I not prepare? First, I just signed up and left home. Second, preparation was not important to me at the time. When I was young, I took life as it came. But Marine Corps Boot Camp was definitely something I should have been proactive about and not reactive. Yes, I ended up graduating Boot Camp with my original platoon and went on to serve our wonderful country for 12 years, 1 month, and 13 days, but my 13 weeks in Boot Camp ended up being much more difficult than it should have been.

The resources available now for those looking to join the military were not around when I enlisted in the Marine Corps. After 12 years in the Marine Corps, and appearing in the *Black Friday: Dark Dawn* series, and spending many more years promoting service of this country, I have often been asked how to prepare for the military. My hope is that you will find some answers in these pages.

2 FIRST STEPS WHEN JOINING THE MILITARY

Generally, the military acquires civilians in recruiting offices located throughout the United States. If you plan to enlist in the military, then your first stop is at one of these offices. Enlisted recruiters are the most visible to the public. If you have seen a military recruiter in places such as a shopping center, you were most likely looking at an enlisted recruiter.

If your goal is to become an officer, then your main options are to apply to one of the military academies (think West Point and Annapolis), enroll in the Reserve Officers' Training Corps (ROTC) program (located at various colleges throughout the country), or head off to Officer Candidate School (OCS).

While the academies and ROTC have their own specialized methods of accepting applications, including requirements that applicants be nominated by someone such as a United States Senator or the President, acceptance to OCS is a more straightforward process that also involves military recruiters. Do not approach the enlisted recruiter if you want to apply to OCS. You need to talk to an officer recruiter. Officer recruiters tend to be less visible to the public, so you may have to expend some effort to find them.

Before walking into a recruiting office, you should really think through your personal goals for your future in the military. What branch of service do you want to join? The experience of serving in the Marine

Corps is different than serving in the Air Force. What environment would suit your personality best? You get to work with technology in the Navy. What job do you envision doing? Being an Army Ranger means getting into top physical shape. There are a lot of factors to consider before you approach the recruiter and you have to do your homework.

Military recruiters are rewarded for achieving particular goals, which are not always aligned with your goals. Recruiters are compensated based on the number of individuals they sign up. Moreover, certain jobs in the military may need to be filled more quickly than others. So the recruiter will do her best to steer you into those jobs. Remember that the job of a recruiter is to fill open positions, and there is nothing wrong with a recruiter doing her job very well. To avoid disappointment, you must acquire as much information as possible prior to walking into a recruiter's office. Know what you want and get all promises on paper.

The same rule holds true when you deal with an officer recruiter. The officer recruiter's job is to get you to sign a contract in order earn a commission and to fill vacancies based on the needs of the military. If those vacancies do not line up with your preferences, the recruiter will work to convince you otherwise.

Consider these negotiations with your recruiter as the military's first test of your toughness. If you are being offered a job that you simply cannot see yourself doing in the military, now is your chance to walk away. Once you sign on the dotted line of the contract, you are committing yourself to the process. So be sure that you are ready.

Why Do You Want To Join?

Before embarking on an experience that will turn your life upside down, it is a good idea to ask yourself "why do you want to join the military?" Without a definitive motivation, it might not be possible to make it through the difficult periods of your military career. The military cannot afford to have anyone who is not entirely committed. Many of the lifelong professionals will find it difficult to work with you if you really have no desire to be there.

There are valid reasons for joining the military, which include a strong desire to serve the country, an interest in the profession, development of teamwork and management skills, and a sense of adventure. There are also questionable reasons that should make you think twice about joining, such as seeking combat opportunities, trying to please others, and "doing something" while you figure out what you really want in life.

The military is looking for level-headed competent operators to join. The job is more "nuts and bolts" than what is portrayed on television. You are as likely to find yourself behind a desk filling out paperwork as being in the field performing operations. The military will expect you to be equally competent as an administrator as you are as an operator. The military has a tendency to weed out the guerilla wannabe and individuals with non-teamwork attributes. When assessing your motivation for serving, it is best to dispose of the conceptions that are depicted on television and in the movies. Individuals who can be trusted with millions of dollars in equipment are far more likely to succeed than those who view military service as something Hollywood.

Initial Paperwork

When you first sit down with your recruiter, she will ask you for various information, such as your full name, social security number, age, date and place of birth, current contact information, information on dependents, current physical fitness regimen, education information, extracurricular activities, past employment history, and prior military service (if any).

Your recruiter may also discuss the best dates available for you to formally begin recruit training at Boot Camp or OCS. This date is very important and should be carefully discussed with your recruiter. You want to arrive at recruit training in the best physical shape possible. You also need to have your mind in order, and your personal life settled. Joining the military is a stress test. Once it begins, they will throw everything at you. But the chances of you failing decrease dramatically if you arrive in the best physical, mental, and emotional shape.

Expect questions from your recruiter that dig deeper into your motivations and history. You may be asked why you want to join the military and why you want to pursue a particular profession. From the moment you walk into the recruiting office, your character is being evaluated.

Character is hands down the most important assessment that will be made, not only prior to your entrance into the military, but all through your entire career. The backbone of the armed forces is trust. If people are going to put their lives in your hands, as will inevitably be the case in the military, they have to trust you. They must have faith that if you come under enemy fire, you have the courage not to cut-and-run. On routine days, trust in your character will matter as well. This means that you have the honor to admit mistakes, and that you do not lie, cheat, or steal.

Armed Services Vocational Aptitude Battery (ASVAB)

If you are enlisting, you will be required to take a multiple choice exam known as the ASVAB. The purpose of the ASVAB is to determine if you have the basic quantitative skills to enlist. The ASVAB is a computerized examination that tests math reasoning, word knowledge, reading comprehension, science, electronics and mechanics (to name a few).

The higher you achieve on the ASVAB, the more career options are available for you in the military. Study for this exam. You can find many resources online, from your local recruiter, or with commercial test preparation software.

Security Background Investigation

One of the most time consuming aspects of joining the military is the security background investigation. The extent of the security background investigation depends on your eventual military job and your needed security clearance.

The military will use the answers you provide in the "Questionnaire for National Security Positions" form to determine your eligibility for a security clearance. The security clearance includes Secret, Top Secret, and Compartmentalized. You are granted access to more sensitive information (on a need-to-know basis) if you have been granted higher levels of security clearance. The greater the security clearance required the more the military digs into your past activities. Some background investigations even lead to a personal interview with an investigator and a polygraph test.

Be forthright in completing all the information on the form. In addition to asking for typical identifying information, including employment and school history, the background investigation may look into the activities of your relatives, your places of residency throughout your life, foreign travel, activities and business interests, and your personal finances.

Why do your personal finances impact your ability to obtain a security clearance? The short answer is blackmail. If you are experiencing financial difficulties, you are considered a higher risk to compromise national security in order to settle your personal debts. Many turncoats have stolen and sold secrets to enemies in order to pay bills.

If you are in heavy debt, start paying them off. Take whatever proper and legal steps necessary to clean up your financial record to show the investigators that you are not a security risk.

Part of your background investigation will also require you to provide personal references. Investigators may be sent out to interview these people, so make sure you carefully choose your character references.

Military Entrance Processing Station (MEPS)

MEPs are large facilities located throughout the United States that are used to intake military personnel. At MEPS, physicians will test your vision, hearing, blood pressure, balance and muscle control. You may be given an electrocardiogram to test your heart. Additionally, blood will be drawn for HIV and cholesterol. You will also be given a drug and alcohol screening and have your fingerprints taken.

Do not be surprised if MEPS feels like a cattle call. A lot of people get processed through MEPS each day and the experience resembles an assembly line. You should also note that some of what happens at

MEPS will be repeated during the first few days of Boot Camp, when your class goes through its initial medical screening.

Certain specialties will require additional testing. For example, if you want to be a military diver, you must complete a pressure test to determine if you can handle extended periods working underwater. If you want to be a pilot, you need to be measured to see if you can fit into a cockpit.

The Oath Of Allegiance

At some point at MEPS, if everything goes accordingly, you will be offered a contract and a military specialty. This is your last chance to walk away. If you agree to the offer, then you sign your contract and you are officially committed to active military duty.

You will then swear in with the Oath of Allegiance. The staff will take you in a room, make you raise your right hand and say the Oath. The Oath is the same throughout the military and each time you are promoted, you take the Oath again. In my 12 years of service, I had the honor of raising my right hand and taking the Oath four times. The Oath is as follows:

"I ________________, do solemnly swear (or affirm) that I will support and defend the Constitution of the United States against all enemies, foreign and domestic; that I will bear true faith and allegiance to the same; and that I will obey the orders of the President of the United States and the orders of the officers appointed over me, according to the regulations and the Uniform Code of Military Justice. So help me God."

Delayed Entry Program (DEP)

Once you are sworn in, you begin the waiting period before shipping off to Boot Camp or OCS. If you enlisted, you enter into the Delayed Entry Program. DEP was established to help men and women prepare for Boot Camp while they are still at their recruiting station.

You need to prepare yourself both physically and mentally to handle the challenges that will be thrown at you. When you go to Boot Camp, you will have Drill Instructors or Drill Sergeants ready to unleash an all new level of hell that most people in the United States never experience. Boot Camp is going to be a challenging endeavor. You must be ready in order to succeed.

DEP exposes future service members to our military way of life. It places them under the direct supervision of active duty enlisted persons at recruiting stations, who have years of experience, training and knowledge. In the Marine Corps, when you are in DEP, you are known as a "Poolee".

While in DEP, you will start to develop a sense of brotherhood – Band of Brothers - through physical fitness, Poolee functions, family nights and other organizational events that develop and build the camaraderie needed to prepare for Boot Camp. Time spent at these events will help you adjust to the military way of life.

It is important to recognize that when you are in DEP, you are under the chain of command of the military. Poolee functions are not "optional attendance" at your leisure, you are under military obligation like any other active duty service member. When you become part of DEP, you sign an agreement to voluntarily undertake military obligations. Your obligations include attending all Poolee functions, contacting your recruiter every week, exercising regularly, reporting any involvement with law enforcement, getting your finances squared away, and putting your personal affairs in order.

DEP transitions you from pure civilian life and exposes you to the initial stages of the military. The purpose of DEP is for you to get all your civilian affairs in order. The day you step on that airplane and fly off to Boot Camp, you must be ready and have no distractions.

3 PREPARE FOR RECRUIT TRAINING

Congratulations. You have taken the first step. Walk Proud.

Your recruiters understand that they need to help you get prepared for Boot Camp or OCS as much as possible. However, they are only going to give back to you according to the attitude you display towards them. Regardless of your background, I would suggest you do the following when interacting with your recruiter:

- Always put your hands behind your back when speaking to them. This shows them respect.

- Always respond to them smartly. That means Yes Sir / Ma'am / Sergeant / Petty Officer. Get rid of slang words.

- Start working on putting yourself in the third person. In recruit training, you will not speak with the words I, me, my, or mine. You should start referring to yourself as "This Recruit" or "These Recruits". Practice so that speaking in the third person is natural when you get to Boot Camp.

- Be willing to step up when recruiters need help. There is nothing better than a recruit who takes care of business and can be depended upon to help get things done. Example: you have PT scheduled at 1630 and the recruiters may be busy. Get the Poolees outside and ready for PT.

- Maintain a good attitude. Humble yourself early and the transition will be much easier when you are put under the pressure of a Drill Instructor or Drill Sergeant.

In Boot Camp, the hours of operation will generally be from 0530 to 2130 (9:30pm) Monday through Saturday and from 0600 to 2130 (9:30pm) on Sundays. Instructors will be with you 24 hours per day, seven days a week while in training.

This will be tough for many of you in the beginning, but just like anything else, your body will get used to it. During the 90 days prior to Boot Camp, try to get into the habit of waking yourself up early in the morning and do so consistently. Buddy up with someone else so that you can be accountable to each other. I would suggest that you each call to ensure the other is awake.

- Monday to Saturday: 0530 – 2130 NO SLEEP
2130 – 0515 LIGHTS OUT

- Sunday: 0600 – 2130 NO SLEEP
2130 – 0545 LIGHTS OUT

Physical Training

Success in the military requires you to be in good physical shape. As such, you should really focus on becoming proficient in certain exercises.

Running

The physical training staple of the military is running. If you do not have the stamina to run the distances required of you in Boot Camp, your chances of success drops significantly. There are volumes of running books on the market. If you are a complete novice, it may be a good idea to purchase one of these books in order to get background knowledge of stretching, proper footwear, rest, shin-splints, pacing, running surfaces, form, and the other important facts necessary to become a competent runner.

Pushups

The second most important exercise skill in the military after running is the ability to perform pushups. A pushup is the ultimate upper-body strengthener. A major problem with pushups is the temptation to cheat with form, including not keeping your back straight, not touching your chest to the ground on the down movement and not fully extending your arms in the up position. When training for Boot Camp, you want to aim for perfect form in all your exercises, as proper form will build up your muscle strength fastest. A correct pushup starts with your hands on the ground a little more than shoulder width apart, arms fully extended, your legs-back-shoulders in a straight line, head tilted up and eyes looking straight ahead. Keeping your legs-back-shoulders straight, bend your elbows until your chest touches the ground, then straighten your arms again until fully extended.

Sit-Ups

For sit-ups, lay on the ground with your knees bent at a 90 degree angle and your feet flat on the floor. Place your fingertips to your ears and place your arms flat to the ground. Curl up with your stomach muscles so that your chin touches your knees. Use no other muscles other than your stomach muscles, and do not assist your upward motion with your arms. Remember to keep your feet planted on the ground as you curl up. When your chin touches your knees, hold for a two-count and then lower your head back down to the starting position. Once you are back in the starting position, do not rest, but immediately proceed to the next sit-up.

Pull-Ups

When doing pull-ups, the palms of your hands must face away from you. The pull-up bar should be perpendicular to the length of your body above your head. Grab the bar with your hands (palms facing away from you), and hang fully extended from the bar. With a smooth motion, pull yourself upward so that your chin is above the bar. Hold this position for a two-count and then slowly lower yourself back to the starting position. Your goal is to perform each pull-up smoothly and with no kicking or kipping to assist your motion. You should only use arm and chest muscles to move up and down.

Strength Training & Swimming

The military's focus on pushups, sit-ups, pull-ups, and running means that strength training and swimming is discretionary for you (except if you are trying out for Special Forces, in which case swimming is extremely important, and aerobic weightlifting does not hurt).

It is beneficial to become proficient in the water even if your ultimate military profession does not involve swimming. Pool-work tones your muscles in very different ways than any other exercise and the dexterity and coordination developed in the water can translate to tangible benefits in other areas of your military career. You will also build lung capacity, which can help with your running and other aerobic performance.

When you are weightlifting, you have the option of using heavy weights and performing few repetitions or using medium weights and performing more repetitions. Fewer repetitions with heavier weights build bulk whereas more repetitions with lighter weights build endurance. From a military training perspective, most of your weightlifting should be done with endurance in mind. That means you should spend the majority of your weightlifting sessions performing more repetitions with lighter weights, and only the occasional session with heavy weights. If you decide to pursue a weightlifting program, there are many commercially available books that teach appropriate technique.

Remember, the way you are taught to do pushups, pull-ups, and sit-ups in this book may be different than the way you are expected to perform these exercises at Boot Camp. The methods outlined previously are designed to maximize your strength and endurance. At Boot Camp, carefully follow the instructions provided for performing exercises.

You will undergo several physical fitness tests while you are in Boot Camp. In order for you to progress to the next level of training, you will have to pass each test.

I cannot speak about the physical fitness tests in the Army, Air Force, Navy or Coast Guard, but in the Marine Corps, the following physical fitness tests are your challenge.

1. *Initial Strength Test (IST)*

The Initial Strength Test is conducted during the first week of Boot Camp, usually the first Friday. The test evaluates a recruit's physical ability upon arrival. The minimum standards for the IST change from time-to-time, but are generally as follows:

Male Recruits

- 2 Pull Ups
- 44 Crunches in two minutes
- 1.5 mile run in 13:30

Female Recruits

- 12 second Flexed-arm Hang
- 44 Crunches in two minutes
- 1.5 mile run in 15:00

2. *Physical Fitness Test (PFT)*

You will receive two physical fitness tests. One will be several weeks into Boot Camp. If you do not pass this test, you will not move on in training. This test does not count as your official PFT but is a requirement to track your progress as an individual and as a platoon. The Final PFT will usually be conducted right before graduation. This gives you time to build your strength and endurance over the course of training. It will consist of 3 events: pull ups, crunches, and a 3 mile run. Each event can result in a maximum score of 100 points for a cumulative maximum of 300 points on the PFT.

The higher your score, the higher your points will be towards your future promotions. The lowest a recruit can score is 135 points total. Trust me, if you do not individually perform during the test, you and your platoon will have a very long day. When I knew that some recruits did not perform during the graded event, we would conduct Incentive Training (IT) on those individuals. If the platoon did poorly, I would take them all to the pit for IT. It can become the longest day of your life if you do not perform to your maximum. Ensure you push yourself because each pull-up, crunch, and second off your run counts for you and your platoon.

The minimum standards for the PFT also change from time-to-time, but are generally as follows:

<u>Male Recruits</u>

- 3 Pull ups
- 50 Crunches in two minutes
- 3.0 mile run in 28:00

<u>Female Recruits</u>

- 15 second Flexed-arm Hang
- 50 Crunches in two minutes
- 3.0 miles run in 31:00

3. Combat Fitness Test (CFT)

The Combat Fitness Test (CFT) will be conducted towards the end of Boot Camp. It is made up of three events:

<u>Movement to Contact</u>. You will start the CFT with an 880 yard run as quickly as possible in the Combat Utility Uniform and boots. In order to pass this event, the males have to complete it in 4:13 or less, and the females a have to finish it in 5:27 or less. Once you have completed the movement to contact event, you will move to the ammunition lift.

<u>Ammunition Lift</u>. You will have to lift a 30 pound (dirt or sand) ammunition can from shoulder level until your arms are completely locked out continuously for two (2) minutes straight. Complete a minimum of 33 repetitions for males and 17 repetitions for females. The more you do, the better your score, so you definitely want to practice in advance and prepare for this graded event. After the ammunition lift, you will move to the maneuver under fire course.

<u>Maneuver Under Fire</u>. The maneuver under fire is a timed, 300 yard event. It will consist of sprinting, low crawl, quick direction changes around cones, casualty evacuation by dragging and fireman's carry, grenade throw, and carrying two 30-pound ammunition cans throughout the course. Males will have to complete this event in 3:58 and females in 5:59.

If you do not pass any of the above physical fitness tests, you could find yourself dropped to the Physical Conditioning Platoon (PCP).

This is where all the recruits who cannot pass the minimum standards are placed. In PCP, your primary focus will be conditioning yourself to get back into training. However, your graduation date will be pushed back and you will not graduate with those who started Boot Camp with you. Some people are in PCP for a week or less and others are there for months. Make sure you prepare physically prior to arriving at Boot Camp, as the last thing you want is to get held back in training. It can be a very depressing place for a lot of recruits because they remain at a standstill while watching their peers move ahead.

Physical training will help your mind, body and soul to become more effective and efficient. The hardest part is to get started. Many people start working out, but after two or three days, they give up and quit.

Just like anything else, you have to develop physical fitness. When done properly, the returns are noteworthy, particularly improvement in muscle tone and overall endurance. On the following pages is a workout program that will start you at the adaptive phase, proceed to the progression phase, and end in the master phase.

Some people like to work out in the morning and some prefer the evening. I would suggest that you work out in the morning because your body has been in an 8 to 12 hour fast (i.e. you have had nothing to eat). If you work out in the evening (which is fine) you will already have eaten breakfast, lunch and possibly dinner.

Adaptive Phase. This is when you start a workout sequence and your body is not used to the program. Your body immediately goes into the adaptive phase to adjust to what you are doing. This phase is difficult because you feel like you are doing everything right and you are not seeing any results. Look at it this way, people did not wake up one day and were 50 pounds overweight, so the transformation is not going to occur overnight. If you want it, this phase needs to be taken very seriously because the harder you push yourself, the faster the excessive weight will leave your body.

Progression Phase. You will find your body getting used to your workout program and you will start to see improvement in tone, endurance and overall appearance. You will build confidence in your program and get to the point where you do not feel right unless you work out. If you miss a day, you will feel sluggish and wish you had worked out.

Master Phase. When you get into the master phase, this means your workouts are not challenging your body any longer. If you do not change your workout, you will stagnate and you will become bored. Though it is easier when you have mastered skills, you slowly start to put your body on cruise control and it will be that much harder to progress if you do not start changing things up.

I am going to give you a 90 day (12 week) workout program that will help you develop your body in preparation for Boot Camp. If you follow this advice, you will be better prepared for the challenges that are ahead. If you find that this program does not push you enough, double it up either by doing workouts twice a day (morning and evening) or double the time during one workout.

This is not the hardest program available, but you must remember, I am trying to prepare you for Boot Camp, not a body building competition or a marathon.

PHYSICAL TRAINING PROGRAM

Week 1 – 3	
Monday	Back & Biceps
Tuesday	3 Mile Run. 3 sets of Max pushups, pull-ups and sit-ups
Wednesday	Chest & Triceps
Thursday	3 Mile Run. 3 sets of Max pushups, pull-ups and sit-ups
Friday	Legs & Shoulders
Saturday	3 Mile Run. 3 sets of Max pushups, pull-ups and sit-ups
Sunday	Off

Week 4	
Monday	3 Mile Run. 3 sets of Max pushups, pull-ups and sit-ups
Tuesday	3 Mile Run. 3 sets of Max pushups, pull-ups and sit-ups
Wednesday	3 Mile Run. 3 sets of Max pushups, pull-ups and sit-ups
Thursday	3 Mile Run. 3 sets of Max pushups, pull-ups and sit-ups
Friday	3 Mile Run. 3 sets of Max pushups, pull-ups and sit-ups
Saturday	3 Mile Run. 3 sets of Max pushups, pull-ups and sit-ups
Sunday	Off

Week 5 - 7	
Monday	Twenty sprints 40 yards long
Tuesday	Legs & Shoulders
Wednesday	Twenty-five sprints 30 yards long
Thursday	Chest & Triceps
Friday	Thirty sprints 20 yards long
Saturday	Back & Biceps
Sunday	Off

Week 8	
Monday	4 Mile Run. 5 sets of Max pushups, pull-ups and sit-ups
Tuesday	4 Mile Run. 5 sets of Max pushups, pull-ups and sit-ups
Wednesday	4 Mile Run. 5 sets of Max pushups, pull-ups and sit-ups
Thursday	4 Mile Run. 5 sets of Max pushups, pull-ups and sit-ups
Friday	4 Mile Run. 5 sets of Max pushups, pull-ups and sit-ups
Saturday	4 Mile Run. 5 sets of Max pushups, pull-ups and sit-ups
Sunday	Off

Week 9 - 11	
Monday	Back & Biceps
Tuesday	400 meter sprints (lap around track) at 80% (4 Times)
Wednesday	Legs & Shoulders
Thursday	400 meter sprints (lap around track) at 85% (6 Times)
Friday	Chest & Triceps
Saturday	400 meter sprints (lap around track) 90% (8 Times)
Sunday	Off

Week 12	
Monday	5 Mile Run. 5 sets of Max pushups, pull-ups and sit-ups
Tuesday	3 Mile Run only
Wednesday	4 Mile Run. 5 sets of Max pushups, pull-ups and sit-ups
Thursday	3 Mile Run only
Friday	4 Mile Run. 5 sets of Max pushups, pull-ups and sit-ups
Saturday	5 Mile Run only
Sunday	Off

This page intentionally left blank

Back & Biceps Workout (NOTE: W=Weight & R=Reps)									
Exercise	Week1	Week2	Week3	Week5	Week6	Week7	Week9	Week10	Week11
Wide Grip Pull-ups									
Dumbbell Curls	W__R__	W__R__	W__R__	W__R__	W__R__	W__R__	W__R__	W__R__	W__R__
Closed Grip Pull-ups									
Seated Concentration Curls	W__R__	W__R__	W__R__	W__R__	W__R__	W__R__	W__R__	W__R__	W__R__
Seated Bent Over Back Fly's	W__R__	W__R__	W__R__	W__R__	W__R__	W__R__	W__R__	W__R__	W__R__
Flat Bar Standup Curls	W__R__	W__R__	W__R__	W__R__	W__R__	W__R__	W__R__	W__R__	W__R__
Standing Back Fly's	W__R__	W__R__	W__R__	W__R__	W__R__	W__R__	W__R__	W__R__	W__R__
Wide Grip Chin Pull-ups									
Lawnmower Pulls w/ Dumbbells	W__R__	W__R__	W__R__	W__R__	W__R__	W__R__	W__R__	W__R__	W__R__
Hammer Curls	W__R__	W__R__	W__R__	W__R__	W__R__	W__R__	W__R__	W__R__	W__R__
Decline Bench Sit-ups									
Sit-up Machine									
Normal Sit-ups									
NAME:			Date Started:						

Chest & Triceps Workout (NOTE: W=Weight & R=Reps)

Exercise	Week1	Week2	Week3	Week5	Week6	Week7	Week9	Week10	Week11
Standard Push-ups									
Overhead Triceps Extensions	W__R__	W__R__	W__R__	W__R__	W__R__	W__R__	W__R__	W__R__	W__R__
Military Pushups (Body Width)									
Chair Dips									
Diamond Push-ups									
Dumbbell Triceps Kickbacks	W__R__	W__R__	W__R__	W__R__	W__R__	W__R__	W__R__	W__R__	W__R__
Wide Hand Push-ups									
Triceps Pull Downs	W__R__	W__R__	W__R__	W__R__	W__R__	W__R__	W__R__	W__R__	W__R__
Decline Push-ups									
Triceps Dips									
Medicine Ball Sit-ups									
Lay Down Leg Lifts									
Hanging Knee Ups									
NAME:		Date Started:							

Legs & Shoulders Workout (NOTE: W=Weight & R=Reps)

Exercise	Week1	Week2	Week3	Week5	Week6	Week7	Week9	Week10	Week11
Standing Squats w/Dumbbells	W__R__	W__R__	W__R__	W__R__	W__R__	W__R__	W__R__	W__R__	W__R__
Standing Dumbbell Press-ups	W__R__	W__R__	W__R__	W__R__	W__R__	W__R__	W__R__	W__R__	W__R__
Front Lunges w/Dumbbells	W__R__	W__R__	W__R__	W__R__	W__R__	W__R__	W__R__	W__R__	W__R__
Standing Side Shoulder Rises	W__R__	W__R__	W__R__	W__R__	W__R__	W__R__	W__R__	W__R__	W__R__
Step Up on Bench Lunges	W__R__	W__R__	W__R__	W__R__	W__R__	W__R__	W__R__	W__R__	W__R__
Seated Dumbbell Press-ups	W__R__	W__R__	W__R__	W__R__	W__R__	W__R__	W__R__	W__R__	W__R__
Walking Lunges w/Dumbbells	W__R__	W__R__	W__R__	W__R__	W__R__	W__R__	W__R__	W__R__	W__R__
Standing Front Dumbbell Raises	W__R__	W__R__	W__R__	W__R__	W__R__	W__R__	W__R__	W__R__	W__R__
Single Leg Squats w/Dumbbells	W__R__	W__R__	W__R__	W__R__	W__R__	W__R__	W__R__	W__R__	W__R__
Seated Flat Bar Shoulder Rises	W__R__	W__R__	W__R__	W__R__	W__R__	W__R__	W__R__	W__R__	W__R__
Seated Leg In & Outs	W__R__	W__R__	W__R__	W__R__	W__R__	W__R__	W__R__	W__R__	W__R__
Alternating Elbow to Knee									
Crunches									
NAME:			Date Started:						

This page intentionally left blank

Diet and Nutrition

This is one of the hardest areas in life to master. America has become a fast food society. We are always on the go and seem to take the easier route when it comes to diet. You can exercise all you want, but if you keep eating poorly, you will not see the results you seek. After my service in the Marine Corps, I found myself still waking up at 0500 and heading off to the gym 5 – 6 times a week. I would work out and do cardio, but for some reason I was still gaining weight.

In the Marine Corps, I was always pushing myself and my Marines. As such, my physical fitness level and weight stayed under control. After leaving the Marine Corps, I was only pushing myself, and my body and my performance in the gym became stagnant. I started to gain some weight. You would think that working out six times a week, one could keep their weight under control. That is not case. I had to truly start looking at what I was eating, figure out what I was doing wrong and start taking the steps to rectify the problems.

It is a myth that if someone exercises long and hard enough, they can eat whatever they want and remain healthy. Granted, there are many examples where it appears that a person gets away with eating poorly, however, your mission is to achieve peak performance during military training. Peak performance will get noticed by your instructors, and more importantly, will make it much easier to deal with the mental challenges thrown your way during Boot Camp.

Even if you are in good cardiovascular shape, if you are carrying an extra five or ten pounds of fat on your body as a result of a poor diet, it is not possible to achieve your peak performance level, and you will feel the effects when you most need energy.

If your diet is currently very unhealthy, it is best not to transform it overnight. Changing your diet "cold turkey" will lead to withdrawal symptoms and cravings, and in short order the temptation to eat junk food and sugar will emerge with a vengeance.

Instead, your goal should be to change one bad habit at a time. For example, if you eat three unhealthy meals a day and cookies in between for snacks, start by swapping out one of your unhealthy snacks for some fruit. That should be your only change for a few weeks until the time when grabbing fruit becomes second nature.

Once that first snack is permanently altered, next consider adjusting breakfast. Substitute unhealthy cereal with a whole-wheat based brand of cereal. Now you have two good habits per day, even though the rest of your diet is unhealthy. Keep this process going, where you swap out one bad habit at a time, until you have reached an optimum diet. This is much more preferable to changing everything at once, which may only work for a few days before you fall back into old habits.

Starvation diets are unhealthy and can lead to medical problems. Instead, start with a normal caloric intake and make small corrections as necessary. As your body gets used to healthy eating and exercise, it should become easier to determine how to adjust as needed.

Here are simple things I did in my life that I believe can help your diet and your life.

Smoking and Dipping. Yes, I said it! I started smoking and dipping when I was 16 years old and did it for over 10 years. I even got to the point where I was smoking and dipping at the same time. The habit grew and I could not get enough of it. I knew it was killing me, but my body wanted more and more. I started making resolutions to quit smoking and dipping. The resolve would last for a while, and then I would have one smoke which then led to two and so on. I wanted to quit, but to no avail. I even got to the point where I would just throw half of a pack of cigarettes or can of dip into the garbage, but would find myself at the store later that day purchasing more.

If you want to quit, the first step is to start speaking about the goal. Example: Every time you see tobacco, you need to say that I am quitting and I do not need it. Even though you might have a dip or cigarette in your hand, start saying what you want to do. Over time, you will start to believe what you are saying. Some of you are thinking, okay, I don't want to quit. So don't. But when you get to that point in life when you are willing, try to quit and start making a better you.

It makes little sense that we go out of our way to take care of our vehicles with the best oil, wax, and soap (i.e. treasuring material things). Treasure yourself and start taking care of your body and it will take care of you. Tobacco is so last year anyway. Being cool is not cool anymore, so get over it.

Sodas and Sweets. The sugar in sodas and sweets is the reason you have a tire around your waistline, dimples on your legs and fat on your butt. Your body holds on to sugar, causing those lumpy areas to get bigger and bigger. Start controlling it! Get over it and move on!

Red Meat. This was the biggest item I recently changed in my diet, and it made all the difference. I have cut all cows out of my diet completely. Why? These animals are full of pesticides, steroids, hormones and antibiotics. Whatever is being pumped into animals is being embedded into the meat and we turn around and eat it.

Okay, so when I decided not to eat cows anymore, it was one of the hardest struggles for me because it seems like everything has cow in it. At first, I really felt like I was starving to death. To make a long story short, I lost about 15 - 20 pounds in three months without changing much of what I had been doing for years. The major change was the meat I was eating.

Did you know that 75% of young men and women are being denied enlistment into the US military? The number two reason is because of a lack of physical fitness or obesity. That should tell you something right? Many people are struggling in this area, so if you find yourself in this category, make the adjustment and it might make the difference you need.

Here are some examples of what you could eat to help your diet. Also, I want you to keep one thing in mind. Drink water. Your body needs about six to eight glasses daily so it can stay hydrated and flush out whatever your body does not need.

Breakfast Examples	Lunch Examples	Dinner Examples
• Egg Whites • Bagels • Pancakes • Oatmeal • Grits • English Muffins • Yogurt • Fruits • Protein Shakes • Whole Grain Cereals	• Salads • Chicken Breast • Turkey • Light Fish • Soups • Tuna Fish • Grilled Chicken • Veggie Burger • Baked Potato • Noodles	• Grilled Fish • Wheat Rice • Wheat Noodles • Vegetables • Chicken Breast • Salads • Baked Potato • Beans • Turkey • Burger w/no Meat

Remember, these are only suggestions. I am sure some of you will get creative with your diets and some of you will not change it at all. At the end of the day, you have to make those decisions for yourself. Just use the above list as a jumpstart and build from it. Tweak it to make your own nutrition decisions, but use the above resources to get started.

If you must have meat, make sure you get it from an organic store where the animals are not pumped full of steroids and pesticides. I suggest that all animal products should come from whole food types of stores, in order to minimize the risk of pumping your body full of poisons it cannot process. Below is a good calculation to help you determine how much food and calories you need to consume daily:

Calculating Resting Metabolic Rate (RMR)

Women: (4.5 x w) + (15 x h) - (5 x a) – 161
Men: (4.5 x w) + (15 x h) - (5 x a) + 5

w = weight in pounds
h = height in inches
a = age

RMR provides an estimate of the baseline caloric intake one needs for daily survival if an individual did no activity.

After calculating RMR, take the resulting value and multiply it by the appropriate number provided on the chart below.

Sedentary Men	1.55	Sedentary Women	1.56
Moderately Active Men	1.78	Moderately Active Women	1.64
Very Active Men	2.1	Very Active Women	1.82

For example, consider a Moderately Active Woman, 68 inches tall, 138 pounds, age 24.

Step 1 (Formula): (4.5 x 138) + (15 x 68) - (5 x 24) – 161 = 1360

Step 2 (Chart): 1360 x 1.64 = 2230

If the above moderately active 24 year old woman consumed less than 2230 calories per day, she will most likely lose weight. If she consumed more than 2230 calories per day, she would likely gain weight.

SAMPLE EATING PLAN

Meal 1	Breakfast 0600	Oatmeal	1 Small Bowl	Juice: Orange Fruit: Banana
		Egg Whites	2	
		Wheat Toast	1 Slice	
WATER				
Meal 2	Snack 0900	Power Bar		Juice: Cranberry Fruit: Apple
		Turkey or Chicken 1/2 Sandwich		
WATER				
Meal 3	Lunch 1200	Chicken Breast	3 Ounces	Juice: Apple Fruit: Strawberries
		Brown Rice	1 Cup	
		Salad	Lt. Oil Dressing	
WATER				
Meal 4	Snack 1500	Power Bar		Juice: Cranberry Fruit: Apple
		Turkey or Chicken 1/2 Sandwich		
WATER				
Meal 5	Dinner 1800	Chicken Breast or Fish		Skim Milk
		Brown Rice/Pasta	1 Cup	
		Potato	1	
		Salad	Lt. Oil Dressing	
WATER				
Meal 6	Snack 2000	Protein Supplement Drink		

Military Knowledge

There are some basic items of information that will keep your Drill Instructors or Drill Sergeants off your back if you know them prior to your arrival at Boot Camp. When I received recruits that knew this information, I felt that those recruits took basic initiative to obtain certain military knowledge. I would consider assigning those recruits to leadership positions within the platoon.

General Orders

Standing watch is a major part of your military career. And in order for you to stand watch, you must know your General Orders. These are the standing instructions that guide every person standing watch throughout the military. Memorize the information below. I also provided scenarios so that you might better understand the General Orders and how they might look on a test you will receive in Boot Camp.

1. To take charge of this post and all government property in view.

2. To walk my post in a military manner, keeping always on the alert and observing everything that takes place within sight or hearing.

3. To report all violations of orders I am instructed to enforce.

4. To repeat all calls from posts more distant from the guardhouse than my own.

5. To quit my post only when properly relieved.

6. To receive, obey, and pass on to the sentry who relieves me, all orders from the Commanding Officer, Command Duty Officer, Officer of the Deck, and all Officers and Petty Officers of the watch only.

7. To talk to no one except in the line of duty.

8. To give the alarm in case of fire or disorder.

9. To call the Officer of the Deck in any case not covered by instructions.

10. To salute all officers and all colors and standards not cased.

11. To be especially watchful at night and during the time for challenging, to challenge all persons on or near my post, and to allow no one to pass without proper authority.

Scenarios

As a sentry, you are a direct representative of the members of your chain of command, and ultimately your commanding officer. You are the senior man onboard your post, unless other ranking members of the chain of command or the commander is present. What general order applies?

1. To take charge of this post and all government property in view.

A sentry is caught sitting down on post, what general order has the sentry violated?

2. To walk my post in a military manner, keeping always on the alert observing everything that takes place within sight or hearing.

A sentry catches a person on restriction leaving the barracks. What general order applies?

3. To report all violations of orders I am instructed to enforce.

A sentry receives a radio message from the Corporal of the Guard. The sentry is told to pass on all information to other guard posts. What general order applies?

4. To repeat all calls from posts more distant from the guardhouse than my own.

A sentry is standing post. The relief sentry is late, and the current sentry leaves his/her post. What general order has been violated?

5. To quit my post only when properly relieved.

The officer of the day tells Post #1 "all sentries will use RED LENS FLASHLIGHTS." What general order applies?

6. To receive, obey, and pass on to the sentry who relieves me, all orders from the Commanding Officer, Officer of the Day, and officers and noncommissioned officers of the guard only.

While standing post, a sentry is telling his buddy about his date last night. What general order has been violated?

7. To talk to no one except in the line of duty.

A sentry sees a fire across the street. What general order applies?

8. To give the alarm in case of fire or disorder.

While walking your post, you come across an unusual incident and you are not sure what to do. What general order applies?

9. To call the Corporal of the Guard in any case not covered by instruction.

The officer of the day approaches a sentry post. What general order applies?

10. To salute all officers, and all colors and standards not cased.

A sergeant walks up to a sentry's post and the sentry states "Halt! Who goes there?" What general order applies?

11. To be especially watchful at night, and during the time for challenging, to challenge all persons on or near my post, and to allow no one to pass without proper authority.

In Boot Camp, there will be eight hours of watch per night. If an officer walks on deck, you will sound ATTENTION ON DECK first, and then you will salute and say the following.

ATTENTION ON DECK!

GOOD (MORNING/AFTERNOON/EVENING) SIR (MA'AM)! RECRUIT _________ REPORTS BLDG _________, SQUADBAY _________ FOR PLATOON _________ ALL SECURE. THERE

ARE ________ RECRUITS ON DECK, AND ________ SERVICE RIFLES, ALL DOUBLE LOCKED, AND ON SAFE. THERE ARE ________ FOOTLOCKERS LOCKED, AND ON ZERO. THE PLATOON IS CURRENTLY ________. THERE IS (SOMETHING/NOTHING) UNUSUAL TO REPORT AT THIS TIME SIR. (State the unusual occurrence, if any).

Do not drop your salute until the officer salutes back and says "carry on" (go back to what you were doing). Then you will cut your salute and say "carry on, aye-aye sir/ma'am."

If it is a Drill Instructor or some other personnel, you will sound ATTENTION ON DECK and say the same greeting, but you will not salute. When you are done saying it all, you will just stand there until you are given the command to "carry on." Once the command is given, you will say "carry on aye-aye sir/ma'am."

You definitely want to make sure you know these procedures like the back of your hand. Just fill in the blanks however you want, but make sure you practice, practice, practice, otherwise Drill Instructors or Drill Sergeants will be all over you, and you will be in for a long stressful night.

In Boot Camp, there are numerous procedures to follow and words that must be spoken in a certain way. As a recruit, the first and last words out of your mouth will be:

> Yes Sir/Ma'am!
> No Sir/Ma'am!
> Aye-Aye Sir/Ma'am!

When you need to speak, you must be at the Position of Attention (POA) and utilize the following procedure:

> Recruit __________ requests permission to speak to Drill Instructor (or Senior Drill Instructor) Sergeant / Staff Sergeant (or Gunnery Sergeant) __________ sir! (or ma'am!)
>
> The Drill Instructor will say "what do you want?"

> You will respond "This recruit requests permission to __________ sir/ma'am."

When you are called to the Duty Hut (Drill Instructors Office), you will bang on the hatch 3 times and say:

> Recruit __________ reporting as ordered sir/ma'am.

When a recruit is called by a Drill Instructor in the squad bay, the entire platoon will repeat what the Drill Instructor says: "Recruit ________________ aye-aye sir/ma'am." The individual recruit will say "Aye sir/ma'am. Aye recruits. Carry on recruits." This may vary somewhat based upon company standard operating procedures. This procedure lets the Drill Instructor know that you have received his/her order and that you are on your way.

Try not to be called out by a Drill Instructor too often in the negative. You want to make sure that if you are getting called, it is for a good reason, not BAD!

4 MARINE CORPS KNOWLEDGE

"Some people spend an entire lifetime wondering if they made a difference in the world. But, the Marines don't have that problem.

--President Ronald Reagan--

Every branch of the service has its own information that you will be required to know and memorize. I am providing information specific to the Marine Corps in this chapter, because it is a topic I can write about with firsthand knowledge. If you are joining the Air Force, Army, Navy or Coast Guard, ask your recruiter to provide you study sheets with all the memorization information you need to know. You should read this chapter even if you are not joining the Marine Corps. This information is very useful, and much of it is similar to what you will learn in the other military branches.

Before we get into the mission of the United States Marine Corps, let me just say right from the beginning, the Marine Corps is not for everyone. Oftentimes young men and women join a particular service out of obligation, not desire. For example, a parent may have previously served in the Marines and now the next generation feels obligated to do the same. Take the time and effort to perform your due diligence and make the most knowledgeable decision, one that is best for you and your future. If the Marine Corps is not right for you, do not pursue that option because of outside influences.

This book is not to steer you towards the Marine Corps. I am simply putting my years of experience on paper to help men and women make a more knowledgeable decision, get a better understanding, and mentally, physically and emotionally prepare for Boot Camp or OCS.

Mission of the Marine Corps

The United States Marine Corps (USMC) is a department of the Navy. The mission of the Marine Corps is to be the amphibious "Force in Readiness" branch of service that is "The First to Fight." The Marines continue to be America's and her allies' emergency force. The mission is simple: Marines are trained and equipped with tactics, techniques and gear to place them on the battlefield and to meet the enemy face to face. The official mission of the Marine Corps was established in the National Security Act of 1947. This was amended in 1952 to make the Marine Corps a "Force in Readiness."

The Marine Corps is one of seven uniformed services. It has continuously proven its war fighting experience by working hand in hand with the Navy in areas of training, logistics and transportation. The Marine Corps is brought together by the Marine Air Ground Task Force (MAGTF), which consists of four elements: Command, Air Combat, Ground Combat, and Support. The Marine Expeditionary Unit (MEU), the Marine Expeditionary Battalion (MEB), and the Marine Expeditionary Force (MEF) are the three types of MAGTF's that contribute towards "Force in Readiness."

The Marine Corps has been and will continue to be an effective and efficient organization because of its willingness to adapt and overcome adversity. The battlefield is endlessly changing, but the Marines have learned to change with it.

History of the Marine Corps

The United States Marine Corps was established on November 10th, 1775 by Captain Samuel Nicholas in Tun Tavern, Philadelphia. Two battalions of Continental Marines were formed. From that point forward, the Marines have fought in every American conflict. Congress codified the Marine Corps in 1798. In the 20th Century, the Marines dominated amphibious war fighting. Marines have effectively responded to crises around the world and the Marine Corps has evolved into an unrivaled force on and off the battlefield.

Over the course of history, the Marines have developed certain characteristics that distinguish them. In 1868, the Marines adopted the Eagle, Globe, and Anchor by order of Brigadier General Jacob Zealin, who was the 7th Commandant of the Marine Corps. The Eagle stands for our nation, the Globe stands for worldwide service, and the Anchor stands for the naval traditions we utilize within the Corps. The term "Leatherneck" was adopted from the time when Marines wore leather straps around their necks to protect themselves from enemy swords. In 1880, the Marines adopted the Latin slogan Semper Fidelis which means "Always Faithful." In 1918 during the battle of Belleau Wood, the Marines were given the nickname "Devil Dogs" for their fierce fighting techniques.

Marines have fought in the Revolutionary War, Quasi-War with France, Barbary Wars, War of 1812, Creek-Seminole Indian War, Mexican War, Civil War, Spanish-American War, Samoa, Boxer Rebellion (China), 1st Nicaragua Campaign, Mexico, Dominican Republic, Haiti, World War I, 2nd Nicaraguan Campaign, World War II, Korean Conflict, Vietnam War, Dominican Republic, Lebanon, Grenada, Desert Storm, Panama, Somalia, Afghanistan and Iraq, to name a few.

There are over 200,000 active duty Marines and 40,000 reservists. The Marine Corps is the smallest branch of service but is considered by many to be the most effective and efficient. Of course everyone has their opinion, but let history speak for itself!

Marine Corps Core Values

We all strive to become part of something bigger than ourselves, whether it is church, a football team, school groups, or other organizations. Each group usually has a mission statement with values to uphold. We push toward these values by surrounding ourselves with people who are heading in the same direction. For the Marine Corps, the established Core Values are as follows:

1. *Honor* (Integrity, Responsibility, Accountability)
2. *Courage* (The right thing, right way, for right reasons)
3. *Commitment* (Devotion to the Corps and fellow Marines)

The primary mission of Boot Camp is to prepare recruits to become basically trained Marines. To become a Marine, you will have to develop Honor, Courage, and Commitment.

Honor. This is the bedrock of our character. The quality that guides Marines to exemplify the ultimate in ethical and moral behavior; never to lie, cheat, or steal, and also to abide by an uncompromising code of integrity; to respect human dignity; and to have respect and concern for each other. The quality of maturity, dedication, trust, and dependability, that commits Marines to act responsibly; to be accountable for their actions; to fulfill their obligations; and to hold others accountable for their actions.

Courage. The heart of our Core Values. Courage is the mental, moral, and physical strength ingrained in Marines. This will carry them through the challenges of combat and allow them to overcome and master any fear; to do what is right; to adhere to a higher standard of personal conduct; to lead by example and to make tough decisions under stress and pressure. It is the inner strength that enables a Marine to take the extra step to do what is right.

Commitment. The spirit of determination and dedication within the Marine Corps that leads to professionalism and mastery of the art of war. It leads to the highest order and code of discipline for the unit and oneself, and it is the ingredient that enables 24-hour dedication to the Marine Corps and country. Additionally it is the proud and unrelenting determination to achieve a standard of excellence in every endeavor.

Commitment is the value that established the Marine as the warrior and citizen who others strive to emulate.

Marine Expectations

- Obey rules and regulations
- Lead by example through their actions
- Respect themselves and others
- Maintain a high standard of discipline and integrity
- Support and defend the constitution of the United States
- Uphold special trust and confidence that others can depend on
- Place faith and honor above all else as a Marine
- Honor fellow Marines, The Corps, Country, Family and Themselves

Leadership Principles

1. Be technically and tactically proficient.
2. Know yourself and seek self-improvement.
3. Know your Marines and look out for their welfare.
4. Keep your Marines informed.
5. Set the example.
6. Insure that a task is understood, supervised and accomplished.
7. Train your Marines as a team.
8. Make sound and timely decisions.
9. Develop a sense of responsibility among subordinates.
10. Employ your command in accordance with its capabilities.
11. Seek responsibility and take responsibility for your actions.

Leadership Traits "JJ DID TIE BUCKLE"

Justice. The quality of being impartial and consistent in exercising command.

Judgment. Ability to logically weigh facts and possible solutions on which to base sound decisions.

Decisiveness. Ability to reach decisions promptly and to announce them in a clear, forceful manner.

Integrity. Uprightness of character and soundness of moral principle, absolute truthfulness and honesty.

Dependability. The certainty of the proper performance of duty.

Tact. The ability to deal with others without creating offense.

Initiative. Seeing what has to be done and commencing a course of action, even in the absence of orders.

Endurance. The mental and physical stamina measured by the ability to stand pain, fatigue, distress and hardship.

Bearing. Creating a favorable impression in carriage, appearance and personal conduct at all times.

Unselfishness. Avoidance of providing for one's comfort and personal advancement at the expense of others.

Courage. A mental quality that recognizes fear of danger or criticism but enables a Marine to proceed in the face of it with calmness and firmness.

Knowledge. Acquire information, including professional knowledge and an understanding of your Marines.

Loyalty. Faithfulness to country, Corps, unit and to your seniors and subordinates.

Enthusiasm. The display of sincere interest and exuberance in the performance of duty.

Leadership is something most of us strive for and want to develop. It is the sum of those qualities of intellect, human understanding and moral character that enables a person to guide a group of people successfully. Applying Leadership Traits is of greatest importance in obtaining confidence, respect, obedience and loyal cooperation. Certain Leadership Traits will be stronger in some and weaker in others. Take the ones that you are weak at, work and develop them to become stronger within yourself. If you do, you will see the relationships you develop with your friends, family, fellow recruits, and Marines improve and you will become more effective and efficient as a leader.

Marine Corps Enlisted Rank Structure

1.	PRIVATE	No Chevron
2.	PRIVATE FIRST CLASS	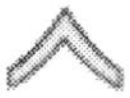
3.	LANCE CORPORAL	
4.	CORPORAL	
5.	SERGEANT	
6.	STAFF SERGEANT	
7.	GUNNERY SERGEANT	
8.	FIRST SERGEANT / MASTER SERGEANT	 
9.	SERGEANT MAJOR / MASTER GUNNERY SERGEANT	 
10.	SERGEANT MAJOR OF THE MARINE CORPS	

Marine Corps Warrant Officer Rank Structure

1.	WARRANT OFFICER 1	 (G)(R)
2.	CHIEF WARRANT OFFICER 2	 (G)(R)
3.	CHIEF WARRANT OFFICER 3	 (S)(R)
4.	CHIEF WARRANT OFFICER 4	 (S)(R)
5.	CHIEF WARRANT OFFICER 5	 (S)(R)

(G) = Gold
(S) = Silver
(R) = Red

Marine Corps Officer Rank Structure

1.	SECOND LIEUTENANT	 (G)
2.	FIRST LIEUTENTANT	 (S)
3.	CAPTAIN	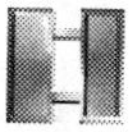 (S)
4.	MAJOR	 (G)
5.	LIEUTENANT COLONEL	 (S)
6.	COLONEL	 (S)
7.	BRIGADIER GENERAL	 (S)
8.	MAJOR GENERAL	 (S)
9.	LIEUTENANT GENERAL	 (S)
10.	GENERAL	 (S)

(G) = Gold
(S) = Silver

For all of the Marine Corps ranks in the previous charts, a recruit will be expected to identify any one of them just by looking at the insignia.

The Marine's Prayer

Almighty Father, whose command is over all and whose love never fails, make me aware of Thy presence and obedient to Thy will. Keep me true to my best self, guarding me against dishonesty in purpose and deed and helping me live so that I can face my fellow Marines, my loved ones, and Thee without shame or fear. Protect my family. Give me the will to do the work of a Marine and to accept my share of responsibilities with vigor and enthusiasm. Grant me the courage to be proficient in my daily performance. Keep me loyal and faithful to my superiors and to the duties my country and The Marine Corps have entrusted to me. Make me considerate of those committed to my leadership. Help me to wear my uniform with dignity, and let it remind me daily of the traditions which I must uphold.

If I am inclined to doubt, steady my faith; if I am tempted, make me strong to resist; if I should miss the mark, give me the courage to try again.

Guide me with the light of truth, and grant me wisdom by which I may understand the answer to my prayer.

AMEN!

5 SURVIVING BOOT CAMP

Boot Camp and OCS are pressure cooker environments designed to do one thing: break you into the military and weed out those unable to handle the pressure. I have one piece of advice for you.

You better want it.

If you do not want the military life with every fiber in your body, your Drill Instructors and Drill Sergeants will figure it out very quickly.

Your mind and body need to be ready for the challenge. If you have outstanding personal issues to resolve, you should take care of those problems prior to your arrival at Boot Camp, because you do not want anything to distract you once you are there. Expect to be completely isolated at Boot Camp until you are granted more privileges that allow you to communicate with the outside world.

As you approach your departure date, you may find yourself having less patience for anyone who tries to waste your time, particularly if you expended significant effort training and preparing for this adventure. You may also have conflicted emotions if you have never left home before. Even simple occurrences like being stuck in traffic, knowing that each minute is one less minute of time spent with your friends or family, will start to bother you much more. Anxiety increases, sometimes to crushing levels. You only leave home for the first time once in your life, and if you are leaving home to join the military, it is one of those experiences that you never forget. You may look back years later and find yourself unable to re-create that anxious feeling of the unknown.

Family members may have a difficult time adjusting to your decision to join the military, particularly those who are opposed to military service. On the other hand, that veteran in your family may be very excited now that you are joining the club. Regardless, it is best to spend some quiet time with your friends and family prior to your departure. If you have any long-standing issues with family members that you want to resolve, now is the time to do it. You need to be mentally free and clear and not distracted in order to have a good chance of succeeding at Boot Camp. The military lifestyle is of such that you could be on long deployments or away from home for extended periods of time. You do not want to leave home with unresolved issues.

You might go to bed with a sinking, pit-of-your-stomach feeling that you get only a handful of times in life. Emotionally you may sense that no matter what happens, you will never be the same again. If you make it through Boot Camp, your first time home will be as a completely different person.

The night before your departure, you should be anxious and may get precious little sleep. Do not let that bother you. You could wake up the next morning with significant apprehension, but also with the mental peace of mind that you are ready for whatever is ahead. However, your brain will play tricks on you. It will try to talk you out of going forward.

But something should keep you pushing forward. You may hesitate to board the final flight to Boot Camp. As the plane leaves the airport and turns for the runway, your heart might beat uncontrollably. Nerves and your own psyche can do crazy things to you. Sadness at leaving home and fear of the unknown will be two dominant emotions. But as the plane takes off and settles to cruising speed and altitude, you should find that your heart rate and anxiety settle down during the flight. There is little turning back at this point anyway.

When you arrive at your destination, the change in climate may catch you off guard. If you are from a part of the United States that is cool or dry, arriving to places like Parris Island, South Carolina or Fort Benning, Georgia in the middle of the summer may be a shock to your system. If you are from a southern state, going to Navy Boot Camp in Great Lakes, Illinois or OCS in Newport, Rhode Island will require some adjustment. It may take your body some time to deal with the change in climate.

Arrival (Receiving)

So it is getting late and you have been full of anxiety all day. On the airplane, you keep running things through your head about what you are getting yourself into and why. While you are doing that, there are some Drill Instructors and bus drivers preparing to head to the airport to meet each and every one of you when you arrive.

Usually recruits come in on one of the last flights of the day. I remember myself back in February 1995, I was on the airplane off to San Diego from New Orleans, Louisiana with my older brother Boe. We had never been on an airplane before in our lives, but I was excited to fly. So, we land in San Diego and we just knew things were getting ready to change forever.

I remember out of the blue there was a Drill Instructor standing down the hallway and immediately he started barking orders to step it out and hurry the freak up. Well, we got in line outside with a bunch of recruits waiting to get on a white bus. I could hear in the background the Drill Instructor just tearing into people. From the point you first meet the Drill Instructors at the airport, expect to maintain a "thousand yard stare", which means that you no longer get to look around.

Once you arrive on base, you will be disorientated with your surroundings. If you have never been on the base before, you may lose perspective as to the size and scope of the area. Drill Instructors take advantage of your lack of orientation to make you feel isolated. They compound this lost feeling by taking the long, winding route when marching your platoon from point A to point B. All of these tricks are used to keep you disoriented.

When you arrive on base, the nervousness that has been gnawing at your stomach may morph into full-fledged anxiety, a feeling you might not shake for several weeks. The unfamiliar military buildings give a foreboding sense of being watched. No matter how strong the anxiety becomes, always remember that many people before you have experienced the same stress and have successfully completed Boot Camp.

I remember stepping off the bus and moving as fast as humanly possible to avoid being next to the Drill Instructor. I ran past them and started doing what everyone else was doing, getting on the yellow footprints on the ground (deck). Then, we started receiving safety briefs and orders we were expected to follow. Then off into the building we went.

Check In

Inside, we got into a room where they go through all your possessions and take anything possible that you were not supposed to bring. They start throwing items away or putting them into a box, which you will not see until the end of recruit training. People, including me, brought so many things that were not allowed and it just made things worse. The recruiters will give you a list of things to bring and not bring, follow them to make life a little easier for yourself.

The first five minutes of Boot Camp will be an event you remember forever. You arrive at some pre-designated intake point to be greeted by staff. Take a look around one last time (if you can), because this is the point of no return. Even if you only last one day, once you cross that threshold, life will never feel the same again.

Your opening moments could go something like this:

"DROP YOUR TRASH NOW! YOU WILL MAINTAIN A THOUSAND YARD STARE AT ALL TIMES. FROM NOW ON, THE ONLY WORDS OUT OF YOUR MOUTH ARE YES SIR, NO SIR, AND NO EXCUSE SIR!"

When words like these are shouted at the top of some instructor's lungs, the world completely changes. Your old civilian life, the one you left behind just seconds ago, becomes a distant memory, and this new experience grows all-encompassing. Your stomach tightens even more and the anxiety that has been consuming you for the last few days morphs into full-fledged nausea. A pumping sensation develops in your head and ears as your blood pressure skyrockets, and instantly a headache begins to form. Then fear kicks in, the fear of someone, anyone, noticing that your eyeballs are looking anywhere other than straight ahead. Peripheral vision disappears. Someone could literally be standing inches away without you ever knowing it. You feel sick to your core with this mix of stress, anxiety, and the unknown. And underlying all of these real physical symptoms is pure unadulterated fear.

You will be marched either as a platoon into a receiving area. All activities become rushed and are performed at a high rate of speed, an artificial but effective way to create stress. The first time you enter a military building, you may be overcome with the sterile smell of cleaning agent. That scent confirms your arrival. It smells like nothing else, is never forgotten, and instantly brings back memories of Boot Camp years later.

Everywhere is noise, yelling, and commotion. The slightest inclination to look around will result in someone screaming at you. Then the first swarming begins. Several Drill Instructors start circling you like sharks in the ocean. Your brain is already on overload from the myriad of instructions that were crammed into you from the moment of arrival. You make a mistake. It could be something as small as moving an eyeball in the wrong direction. The Drill Instructors descend on you like bees defending their hive. Each one screams a string of insults and questions inches from your face, oblivious to the demands of the other Drill Instructors and unwilling to wait for an answer. Sensory overload. Your brain panics, and confusion grows with no chance to think ahead. You blurt out "NO SIR!" to one Drill Instructor, only to be yelled at by another for not answering his question with a "YES SIR!" Many victims lose their balance, begin to sway, and sometimes collapse. The Drill Instructors feed off of the fear, building more momentum. Then for no apparent reason the sharks retreat, returning to the hunt, slowly circling for their next prey.

No one can tell how long the swarming lasts, but to the recipient it takes forever. Deeper into the receiving building, you become more disorientated, and the headache grows more painful. All sense of direction is lost, despite having only walked through an entrance and turned into one passageway. The size of the building is irrelevant. When moving, you are required to keep your right shoulder against the wall (bulkhead) at all times, and always end up taking the long route to your destination, adding to the sense of confusion and isolation.

After your belongings have been inspected, you are off to get uniforms, haircuts and briefs. You do NOT get to sleep the first night you arrive because you are doing all kinds of check-in related activities. The next morning at zero dark thirty, they will get you outside on the road to bring you to chow. By this time, you are starving and cannot wait to sit down and eat. I remember the Drill Instructors saying whatever you put on your plate you will eat. The food was nasty the first time because there was no salt or seasoning but you eventually get used to it. Needless to say, I could not eat everything on my plate, but I tried because the last thing I wanted was to get in trouble from the Drill Instructors.

Newly issued military gear distinguishes recent arrivals from more senior recruits. For males, your head is almost 100 percent shaven. Women's hair is cut to a maximum of shoulder length. These indications help Drill Instructors identify you as a newbie and signal to the whole command that it is open season on you.

During marches, your eyes will fixate on the neck of the person directly ahead. Look closely enough and you see veins popping out under intense pressure. Since fear prevents you from looking anywhere else, you quickly become disoriented.

The tension can be cut with a knife, and your brain starts playing even more mind games. For example, each time the platoon files into formation, which usually consists of four columns and an unlimited number of rows, you hope that you end up somewhere in the middle. Anyone buried inside the formation might hide out from the Drill Instructors. These self-inflicted mind games build more fear as you hope for the luck of the draw during each evolution. But inevitably luck runs out, and you are the one arriving first in formation or who marches incorrectly.

Trying to avoid mistakes at all costs, new and horrifying levels of hell are unleashed on fellow recruits, and all you can do is tell yourself, "Please don't make a mistake". Your brain moves into overdrive as it tries to figure out ways to avoid having the Drill Instructors turn all their attention only on to you.

Two matters that will aggravate you are sleep deprivation and procedures put in place solely for annoyance. Neither of these issues can be fully understood until you actually experience them yourself. Whatever few hours of sleep you get each night, you will go to bed tired and wake up exhausted. Each day the sleep deprivation will mentally wear at you just a little bit more.

Without the sleep deprivation, many of Boot Camp's challenges would be easier to handle. This includes annoying procedures that recruits are required to perform, such as knocking on doors a certain way, requesting permission to enter different rooms, or reciting phrases at particular times while using certain voice intonations. Each little procedure you are required to follow at random times throughout the day is put in place to cause frustration and to trip you up. There are so many little aggravations designed into Boot Camp that you will inevitably make an error, and trying to keep track of all the procedures and games will make you even more mentally fatigued.

Settling In

Life starts to settle into a reasonable routine a few weeks into the course. You start getting used to the surroundings. Drill Instructors still try to make your life miserable, but you enter into a mental zone where anything thrown your way bounces off you like Teflon. That is the goal of military training. When nothing bothers you anymore, you can be sent into danger and still perform your duty. You instinctively know when you mentally get to this point. It usually coincides with a new sense of fearlessness and confidence. Once you achieve this mindset, completing Boot Camp becomes a matter of making it through the various inspections, classes, and other training requirements.

Be prepared for one reality: there is never enough time in the day. Boot Camp is purposefully designed so that recruits feel that there is always time pressure. Boot Camp forces recruits to continually evaluate what responsibilities are vital and which ones should take lesser precedence.

Inspection

On ongoing part of military life are inspections. Whether it is your uniform, your equipment or your squad bay, inspections in the military are generally the way for a command to "reset" everything to a new state. Over time, especially when a unit is heavily deployed, equipment and personnel get worn down. Tools go missing. Vehicles get dirty. It is a fact of military life.

To reverse the downward spiral, commands regularly schedule various inspections in order to get everything squared away. It is no different in Boot Camp.

On inspection day, Drill Instructors will scrutinize your squad bay, your uniform, your equipment, and your weapon (if you have been assigned one), searching for the slightest discrepancy. You may be expected to answer military knowledge questions under intense pressure (to test if you lose your cool and experience "brain freeze" when under stress).

Immaculate rooms, uniforms, and equipment are not emphasized during inspections just for the sole purpose of making your life miserable. The military needs service members who have total attention to detail. Small mistakes cost lives. If an important item of equipment is broken when you are in an operational environment, you are expected to catch and rectify the problem.

So in Boot Cap, when you fail an inspection because a button was not properly secured, or clothes were not folded properly, or your sheets were not precisely tucked, this has nothing to do with turning you into the world's greatest housekeeper. If you come away from an inspection thinking that the whole purpose was for you to have a perfect uniform and flawless locker, you are missing the entire point of the exercise. Boot Camp is trying to teach attention to detail, and the most effective method to accomplish this goal is by inspecting uniforms, lockers, barracks, and equipment for minor discrepancies. Recruits who catch a misplaced thread or an unbuttoned shirt pocket today in Boot Camp will be the Marines, Soldiers, Airmen, and Sailors who catch an unsecured fully-loaded firearm or a missing bolt on a jet engine tomorrow (and saving someone's life).

The other purpose of the inspection is to provide so little time to accomplish everything, it is impossible to succeed without working as a team with your fellow recruits. The ability to coalesce as a team is one of the most important attributes of a successful military machine.

Parade and Rifle Drill

Marching in formation is a military staple. Additionally, Boot Camp usually has some version of drill, primarily with recruits utilizing rifles. The purpose of both is to develop cohesion, discipline, teamwork, muscle memory and coordination. During the first few days of Boot Camp, when your platoon moves between various locations, falling into formation and marching around as a team is an unmitigated disaster.

But through repetition and constant practice, eventually recruits adapt and become comfortable marching as a unit and conducting drill. It takes enormous individual discipline to get the team to this point. During drill, if you are capable of not breaking formation when a bee lands on your shoulder, you are less likely to panic years later when you are facing enemy fire and your weapon malfunctions.

The Pit

Boot Camp usually has a particular location, known as a grinder, quarterdeck, pit, or some other name, where individuals or whole platoons are taken for various infractions and are provided a dose of what is colloquially known as "Punishment PT". Drill Instructors want to see who is scared of the grinder. Who tries to avoid the pit at all costs? If a recruit is terrified of a little Punishment PT, why should that person be trusted to handle real pressure situations in a potentially lethal operational setting? Some of the recruits in your platoon will have no fear of the grinder. Your Drill Instructors will respect those who have the "bring it on" mentality (while not being reckless or dangerous).

Drill Instructors like to test cohesiveness in the grinder as well. They may suggest that a particular individual is not putting out enough effort and so the whole class has to do pushups longer. Drill Instructors want to see how others react. Does the rest of the platoon yell at the "slacker"? Fail. Does someone shout out threats? Fail. Or does the platoon offer encouragement? Succeed. The military is team first.

Your goal is to become, or show that you can become, proficient in everything that you are tasked to do. Drill Instructors are looking for confident individuals who are willing to take on any challenge, learn any assignment, and make mistakes but own up. Willingness to take on responsibility is the key. Your Drill Instructors may have 10 years or more of military experience, have seen many professionals in their day, and want their future colleagues to recognize the immense responsibility that being in the military entails.

Taking A Step Back

It takes time to get over the initial fear of Boot Camp. The faster you get over that fear and the sooner you throw yourself into displaying maximum effort, the easier it becomes to survive the training. There are some recruits who live in constant fear while others overcome it within days of arrival. Your goal is to be the latter. Your Drill Instructors will respect you more the faster you toughen up.

Some recruits attempt to hide from their Drill Instructors by not getting noticed at all during Boot Camp, and a few of them pull it off. But doing so defeats the point of Boot Camp, which is to get over your fear. If you spend your entire time in recruit training trying to avoid getting noticed by your Drill Instructors, whatever fear you have will not be confronted. You will end up spending an inordinate amount of time, effort, and mental energy figuring out how not to get noticed and not to get screamed at by your Drill Instructors rather than focusing on the moment and facing your fear down. Remember, your Drill Instructors would prefer to spend their time messing with recruits who are terrified of them. The sooner you get over your fear, the sooner the Drill Instructors focus their energies elsewhere.

Squad Bay

Boot Camp is designed to simulate the type of chaos you will experience many times throughout your military career. Your goal is to focus on the task at hand and the orders you need to follow no matter what kind of pandemonium is going on around you. Sleep deprivation adds to the challenge. That is the whole point. When deployed, you could go through days of activity with no sleep, and you will still be expected to function and perform your duties the same way as if you had just woken up from eight hours of sleep.

While Boot Camp appears to have numerous pointless evolutions, in reality, much of what goes on is put in place in order for you to get used to the military environment. In the thick of the most stressful moments, it is easy to forget that everything at Boot Camp has a purpose. With the limited amount of time available, the military throws everything at you so that you adapt quickly and effectively.

At the end of the day, the most important decision is to Never Quit and Never Give Up! Yes, there are going to be times when you want to say "screw this," but you cannot take the easy path. You have to stay focused on why you decided to join the military. Lean on your fellow recruits to help you through the hard times. The same goes for them. When you are down, they will need to be up. When they are down, you will need to be up. Together both personally and professionally, you will pull through the training and make it through Boot Camp.

Top 10 things you can EXPECT in Boot Camp

1. Expect to Get Yelled At!
2. Expect to Be Exhausted.
3. Expect Long Days and Extremely Short nights.
4. Expect to get Sick (Recruit Crud).
5. Expect Stress.
6. Expect Drama.
7. Expect Drill Instructors in your FACE.
8. Expect the Unexpected.
9. Expect to be Transformed.
10. Expect your life to NEVER BE THE SAME!

Expect to Get Yelled At!

First and foremost, Drill Instructors are known for causing hate and discontent. No matter how good or bad you are, they will somehow or someway find something wrong with you. You are going to either get caught moving slow, looking around, being undisciplined, or doing something wrong. This will cause the Drill Instructors to move into your personal space and unleash on you. When they do come into your space, you must ensure that you maintain your bearing and stay focused. Instructors will tear into you, demand that you answer a bunch of questions, and check to see your level of discipline. If you lose your bearing, they are going to feed on that and continue messing with you until you are held accountable or show the level of professionalism needed to be left alone. For example, if I came into a recruit's personal space and lit them up, if they kept their head and eyeballs straight to the front, did not lose their bearing, and sounded off with Yes Sir!, No Sir!, or Aye-Aye Sir!, I would leave their space shortly thereafter. If you want the Drill Instructors quickly in and out, follow the guidance above. If not, you are going to have a very long day. Regardless, you will get yelled at!

Expect to Be Exhausted

The first night you arrive in Boot Camp, it will be zero dark thirty. Basically, you arrive late at night and will not be sleeping. You have to go through processing right from the beginning, all night and throughout the next day. You will not get to go to sleep until the following night. By that time, you are extremely exhausted and just ready for bed. However, there are always two to four people on Fire Watch, which is "duty" watching over the barracks while other recruits go to sleep. The Fire Watch is usually one hour long where you have to be responsible for everything and everyone.

The short eight hours you get to sleep will definitely not feel like enough. Some recruits will only sleep around six hours because they have to be up prior to their fire watch shift and after their shift. This will immediately cause fatigue. If you are not on Fire Watch the first night, you will end up on the second or third night because of the platoon rotation.

The training throughout the entire cycle (13 weeks) will personally and professionally push you above and beyond your own limitations. That within itself will cause stress, leading to exhaustion. This is a test of your will to survive Boot Camp. Remember, however, the Drill Instructors will be up before and after you. While you are sitting, they are standing. While you are sleeping, they are roaming. While you are resting, they are preparing to be the example of what it is to lead, while you recover from exhaustion.

Expect Long Days and Extremely Short Nights

The training schedule of Boot Camp is from reveille (before sun up) to taps (after sun down). It will continuously feel like one event to another. Most Drill Instructors do not inform the recruits of what is on the training schedule for the day, so it keeps everyone guessing. You will find yourself cleaning, eating, moving, running, marching, drilling, sitting in classes, PTing, and much more throughout the day. Some days will feel like an eternity and others will fly by. Because you are moving from one event to another, when nighttime comes, it seems so short. As soon as you lay your head on the pillow, you get woken up for Fire Watch or reveille. When you have Fire Watch, you will notice that your nights are even shorter, making it very difficult to get up and stand your duty. Just mentally prepare yourself in advance for long days and very short nights.

Expect to Get Sick (Recruit Crud)

Sickness in Boot Camp is very common. No matter how strong your immune system is, you are constantly exposed to germs from past and present recruits. Cycle after cycle, recruits use the same mattresses. You are issued your own sheets, but germs are still prevalent. Many recruits get dropped because they get too sick. Key advice is to ensure you hydrate (drink a lot of water). You are going to get sick, and the more you hydrate the quicker the illness will leave your body. I have seen recruits go to medical never to be heard from again. Yes I was sick when I was a recruit, but I was not going to let that stop me from accomplishing my goal, which was to graduate with the recruits with whom I started.

Expect Stress

The Drill Instructors will place upon you pressure and stress like you have never endured before. Physical, mental, and emotional strain will take you to the point of exhaustion. This pressure helps to break the undisciplined habits developed over the course of your life. The hardship that the Drill Instructors will place on you is designed to push you above and beyond your own limits. The pressure will develop you into smartly disciplined, physically fit, basically trained military personnel. There have been times when recruits are pushed to the point where they want to give up and quit. You cannot let the stress get the best of you because the day will eventually come to an end.

You need to realize that the Drill Instructors will purposefully place undue pressure on you. Continuous training of your body and moving from one event to another without breaks will cause stress. You have to develop the ability to maintain control when dealing with the Drill Instructors, fellow recruits, and the excessive demands being placed on you. Try to see how the other recruits are dealing with the situations that stress you out. It is organized chaos, so you have to find a way to perform to the best of your ability while remaining calm. If not, the

Drill Instructors will zero in on you. They will then put even more stress on you because you are standing out from the norm.

Expect Drama

Drill Instructors are very theatrical and dramatic in their approach to training recruits. That is why you see them moving at the speed of light and throwing their hands up and down as if they lost their minds. Even if you are not looking directly at them, you can see their dramatic reactions. The Drill Instructors will try and cause as much drama as possible. Try to not let your guard down as a recruit. You have to stay focused and disciplined. Remember, it is only stress if you let it stress you out. If you let the drama get to you, your emotions will show and the Drill Instructors will be dramatic on you.

Expect Drill Instructors in your FACE

If you think for one second they are not going to be in your face, screaming at you at the top of their lungs, you have another thing coming. There have been times when I would spend over a minute in recruits' faces to the point they just close their eyes and receive it like a champ. The last thing you want is to lose your discipline when the Drill Instructors are up close and personal. They will definitely break into your comfort zone. The time they spend in your personal space will last as long as you want it to last.

Expect the Unexpected

Most recruits expect training to be a certain way. Their level of prior exposure either helps or hinders their adjustment.

Information from your recruiting station, television, books you have read, and the videos you have watched will form a certain expectation of Boot Camp. There will always be times when you are not expecting things to happen. However, the more you have seen and developed a mindset on how Boot Camp can be, the more you will be able to deal with those unexpected moments that crop up from time to time. Expect things to catch you off guard and when the unexpected happens, you have to efficiently deal with each and every situation that arises. The Drill Instructors will and must catch you off guard often in order to get you to become more effective in your training.

Expect to be Transformed

Transformation does not and will not come overnight. You have to work for it, you have to strive for it, and you have to believe in it. During the training process, you will find your physical, mental, and emotional strength develop and eventually you will feel the transformation happening right before your eyes. The key is to have a very positive attitude and to want to change and be transformed. If you apply yourself both personally and professionally, you will transform. Believe in yourself and your future.

Expect your life to NEVER BE THE SAME

When I joined the Marine Corps, I really had no idea of what I was getting myself into. I just knew that there was something better out there and I had to strive towards my future. The first several days in Boot Camp, I felt completely lost. I felt that I was never going to catch on, but I was willing to do what it took to keep the Drill Instructors off my back. So I would follow what others were doing until I started to catch on to the program. Once I attained a certain level of understanding in Boot Camp, life started to change. I suspect if I would have prepared prior to my arrival, life could have been a lot different and much easier for me.

For the first time in my life, I was being exposed to every race, nationality, ethnicity, and culture. Right from the beginning, regardless of your past mindset, you come to Boot Camp and individuals formerly considered outsiders immediately become part of your daily life. At first you may find yourself somewhat reluctant, but you are forced to focus on working as a team. In the short term, it is great because you keep the Drill Instructors off your back. Then the transformation begins.

It is not even really noticeable, but then you start to question life as you knew it. You become faster and smarter physically, mentally, and emotionally. When graduation arrives, you begin to take pride in yourself like never before. I remember feeling like I could do anything.

I was young, motivated, and ready to serve this wonderful country of ours. As I stood at the position directly in front of my parents, I knew I was different, I felt I was different, and to everyone around me, looked entirely different. I developed military bearing and discipline from all the pain and suffering endured over the course of three months. It is probably one of my most noteworthy accomplishments that I have ever achieved in my life. From that day forward, I was no longer striving to be part of something bigger than myself, I was part of a history and a legacy that has dated back to 1775.

6 MARINE CORPS BOOT CAMP SPECIFICS

All military Boot Camps and Officer Candidate Schools are challenging. The information provided in this book is useful for you whether you join the Army, Navy, Air Force, Coast Guard or Marine Corps. But since I served as a Marine Corps Drill Instructor, I am providing additional information in this chapter specific to the Marine Corps. You should not skip this chapter even if you are joining another service. Remember that this information is useful, and much of it will be similar to what you experience in other military branches.

The schedule for your first week in Marine Corps Boot Camp will look something like this:

TUESDAY	
REV/CHOW	**0400-0530**
MEDICAL 1	**0530-0945**
DENTAL 1	**1000-1300**
RLS	**1315-1530**
URINALYSIS TEST	**1545-1745**
CHOW TIME	**1800-1900**
ADMIN TIME	**1900-2000**
TAPS (BED TIME)	**2000**

WEDNESDAY	
REV/CHOW	0400-0530
MEDICAL 2	0530-1045
DENTAL 2	1100-1630
COMP TIME	1645-1745
CHOW TIME	1800-1845
MEDICAL FORMS	1845-1930
TAPS (BED TIME)	2000

THURSDAY	
REV/CHOW	0400-0530
CLOTHING	0530-0935
COMP TIME	0950-1120
ID CARDS	1135-1300
SMART CARDS	1315-1445
COMP TIME	1500-1745
CHOW TIME	1800-1900
TAPS (BED TIME)	2000

FRIDAY	
REV/CHOW	0400-0530
PFT / IST	0730-0935
COMP TIME	0950-1120
MOVEMENT	1135-1300
BLACK FRIDAY	1300

On Friday, you will be up again around 0400 and off to chow. In the early morning, you will take your official Initial Strength Test (IST). If you do not meet the minimum requirements, you will NOT get to continue training. You will be placed in the Physical Conditioning Platoon (PCP) where you will be set back until you can meet the requirements to continue training.

Male Recruits	Female Recruits
• 2 Pull Ups • 44 Crunches in two minutes • 1.5 mile run in 13:30	• 12 second Flexed-arm Hang • 44 Crunches in two minutes • 1.5 mile run in 15:00

Once you have passed the IST, you will head back to the receiving barracks where you will take a PT shower (basically run in, get wet and get out). Then, you will get all of your belongings packed into your war bag and sea bag. You will be leaving with all of your issued items and making your way down to your training company, where you will begin your training. Warning, this will be one of the longest walks of your life because you will be moving ALL of your gear at one time!

The sea bag and war bag are stuffed to the gills and you have to personally carry them to your new barracks home. Hopefully, your barracks is not that far away. But more than likely, your new home will be quite far, so just beware. You will take your items into your new barracks and the receiving Drill Instructors will get you and your gear pre-staged. Then it is back on the road to chow. Once chow is done, you will head to a classroom where you will meet some of your senior company leaders with whom you will start training.

Soon after, you will be off to meet your assigned Drill Instructors for the first time, the ones who will be training you for the next three months. You will be placed on the floor in a classroom type setting. The series commander will give you a briefing, and then have all of the Drill Instructors come out of the Duty Hut and stand before you. From there, the series commander will have the Drill Instructors reaffirm the Drill Instructor Oath, which is as follows:

"These recruits are entrusted to my care. I will train them to the best of my ability. I will develop them into smartly disciplined, physically fit, basically trained Marines, thoroughly indoctrinated in love of Corps and Country. I will demand of them, and demonstrate by my own example, the highest standards of personal conduct, morality, and professional skill."

The Senior Drill Instructor (SDI) will then take over from there. He will give his introductory SDI Speech. Right from the beginning, I want you to think about the following, because these will be demanded from you by your Drill Instructors:

- You must give 100% of yourself at all times.
- Obey all orders, instantly, willingly, and without question.
- Treat all Marines and recruits with courtesy and respect.
- You will not physically or verbally threaten another Marine or recruit.
- Be completely honest in everything you do.
- A Marine never lies, cheats, or compromises.
- Respect the rights and property of others.
- A Marine never steals.
- You must work hard at strengthening your body.
- Be proud of yourself and the uniform you wear.
- Try your best to learn the things you will be taught.
- Above all else **Never Quit and Never Give Up!**

We are offering you the challenge of recruit training, and the opportunity to earn the title United States Marine! An opportunity is EARNED, NEVER GIVEN! From there, the green belt Drill Instructor will take over and you will be introduced to one of the longest days of your life, known as "pick-up". Marines who have gone through pick-up call it Black Friday because pick-up happens on a Friday and recruits do not remember much of that day due to the ongoing chaos.

Forming

Saturday, Sunday and Monday are considered FORMING Days. It is an opportunity for the Drill Instructor to teach the recruits the basics, items they need to learn right from the beginning. It is vital that the Drill Instructors set a solid foundation on what is expected from each of you.

Below, you will find the In House Procedures checklist that I would print out on a 5 x 8 card. As I completed teaching recruits a particular class, I checked it off the list. There are a lot of items that need to be taught inside the barracks, with little time to get them done. The In House Procedures are usually taught by the Experienced Drill Instructor "EDI" or Junior Drill Instructor "J." These Drill Instructors are right below the Senior Drill Instructor and are responsible for teaching the recruits just about everything during recruit training.

In House Procedures Taught During Forming

Procedure	**Description**
Gear Inventory	Drill Instructors will inventory your personal effects to ensure you have it all
Church Counts	Number of recruits belonging to each religion
Fire Drill	Drills to leave the barracks in case of fire
Rack Making	How to make your bed
Cammie Tag	Put your name on uniforms and set some aside for special events during training
Mark Gear	Labels for your racks, rifle racks, foot lockers and locks
Class Procedures	Setting up the classroom inside the barracks

Procedure	Description
Shoe Displays	Placement of boots, shoes and shower shoes under your rack in one line with laces tucked in
Moonbeam Cover	Mount your flashlight to your rack and hang your cover while inside the squad bay
Designate Squads	Place recruits in their squads
Fire Watch	Procedures for reporting, wake-up and roster
Chow Hall	How to form for chow and how to conduct yourself at the chow hall
Command Repeat	How recruits repeat what Drill Instructors say in order to keep the platoon informed
Alignment	How to stand at the position of attention and where to stand
Foot Locker Setup	Get all items that belong to you organized in your foot locker
Report Procedures	How to report your post (guard)
Platoon Sizing	Get the platoon in height order from tallest to smallest for uniformity of appearance
Clean The House	Designate cleaning areas and how to conduct cleaning properly, also known as field day
DI Highway	The middle of the squad bay, an area where recruits are not allowed
Duty Hut Report	Properly reporting to Drill Instructors
Time On Deck	Last fire watch (0430 – 0530) will give time on deck calls by banging on DI's hatch and saying "SIR/MA'AM, TIME ON DECK IS ________________ SIR/MA'AM!"

Procedure	**Description**
War Belts	Teach recruits how to adjust their war belts on their waists
Rifle Locks	Cables and locks on rifle racks for when your weapons are issued
Billet Holders	Drill Instructors will assign the platoon guide, squad leaders, and other positions
Head Procedures	How to attack the head, making head calls and what to do to properly secure the head
Diet Cards	Issue overweight recruits tags for eating at the chow hall
Road Guards	Assign the recruits who put on reflective vests that alert vehicle traffic of platoon movement
Cut Belts	Sizing your belts down to proper standards
Mark PT Shirts	Platoon number and recruit name stenciled on shirts designated for physical training
Gear Guard	Procedures how to properly stand post
PT Gear Staging	Placement of war belts and canteens when you are on the PT field
Hygiene	Notifying Drill Instructors of hygiene issues
Appearance	How to wear your uniform sharply
Verbal Greetings	Greet everyone at all times
Small Unit Leaders	Recruits in charge, such as the guide, squad leaders, and fire team leaders
Cammie Taco	How to get items such as toothpaste and toothbrush set up for hygiene inspection

Procedure	Description
Monday Cleanup	Assign recruits to morning cleanup crew for company office
Speaking	Proper procedures to keep you out of trouble when speaking with a Drill Instructor "SIR, RECRUIT ________ REQUESTS PERMISSION TO SPEAK TO DRILL INSTRUCTOR / SENIOR DRILL INSTRUCTOR ________ SIR!"
Laundry	How to mark laundry and where to place laundry when dirty
Meds	Medicine turned over to Drill Instructors
Early / Late Chow	Designate the recruits who will watch over the squad bay while the platoon is off to chow
Boot Lacing	Getting your boot properly laced with a bridge
Basic Daily Routine	What you are doing each morning and evening as a recruit
Cadence	Teaching recruits how to properly sound off when cadence is called
Linen Survey	How to fold your linen, get linen turned in, and get linen cleaned weekly
Stacking Weapons	Once your weapons are issued, you will be taught how to stack your weapons properly prior to entering the chow hall
Colors	What to do for morning colors and raising of the American flag
Last Fire Watch	What to do with laundry and how to announce time by banging on Drill Instructor hatch

You might see Drill Instructors with 5 x 8 cards in their hands. Now you know some of the information those cards may have on them. Some Drill Instructors would just write the information on a white board inside the Duty Hut. However it is done, the written information helps Drill Instructors stay focused on what needs to be accomplished.

Everything I have covered so far is to prepare you to officially start your training schedule. By this time, you are very tired, stressed out and wondering what in the world led you to join the Marine Corps. You feel that the Drill Instructors are robots and have no personality. You are correct but this is incomplete information. Marines do not come down to the level of other people. Instead, we bring them up to the leadership standards expected by the Marine Corps. We do so by pushing, motivating and inspiring people. Accountability comes in Incentive Training, which is how Drill Instructors motivate and instill discipline in recruits.

Details On Incentive Training (IT)

When you do something wrong, you are going to pay for your mistake through many different exercises such as:

Side Straddle Hops. Jumping Jacks.

Stationary Double Time In Place. Running in place with high knees.

Pushups. All the way up, all the way down to where your chest touches the deck.

Arm Rotations. Extending your arms out and rotating them big and small.

Mountain Climbers. Push up position and with one leg forward and one leg back and with a slight hop, you bring the back foot forward and front foot backwards.

Steam Engines. Stand with hands behind head, take your left elbow up to touch your right knee and then the other side.

Abdominal Crunches. Laying on your back with your arms crossed over your chest and your feet off the ground (deck) taking your elbows to your thighs.

Rifle Physical Training. Taking your service rifle and conducting ins and outs, above your head and back of your head, forward to upwards.

These exercises are no joke with a Drill Instructor yelling at the top of his or her lungs and pushing you faster than you ever imagined. There is really no way to avoid IT, so you should not fear it, because it will be imposed on individuals and on entire platoons throughout recruit training. In the beginning of recruit training, the Drill Instructors are looking for any and every mistake (looking around, scratching your face, getting an attitude, losing your discipline, not sounding off, displaying no team work, etc.) so they can take you to the quarterdeck or sand pit to make you pay with IT.

If you want IT less than others, make sure you are paying attention to what is going on. The quicker you pick things up, the easier life will be during Boot Camp.

Academic Classes

Marine Corps Common Skills classes will occur throughout recruit training. These classes consist of but are not limited to: Intro to the Service Rifle, USMC Values, Interior Guard, Customs and Courtesies, First Aid, Uniforms, USMC History, Core Values, Courage, USMC Mission, Honor, Code of Conduct, Leadership, Law of Land Warfare, Equal Opportunity, Sexual Harassment, Sexual Assault, Sexual Responsibilities, Substance Abuse, Suicide Prevention, Hazing, Operational Risk Management, Education Benefits, Driver Safety, Mentoring, Travel, Civilian Attire, Personal Appearance, Thrift Savings Plan, Terrorism Awareness, Operational Security, Fraternization, Marriage, Management & Financial Responsibility, Combat Readiness, and Combat Skills.

While you are in classes, there is always a Senior Drill Instructor at the back. They usually will not say anything during class, but immediately after, the Senior Drill Instructor will get the name and platoon number of those who were moving around a lot, scratching their face, not sitting properly, or falling asleep. Those individuals will receive Incentive Training after class.

We all get tired. You will have to get up out of your chair and stand up in the back so that you do not get in trouble. If you see others falling asleep, find a way to get their attention and try your best to keep them from getting the attention of the SDI. There have been times when too many recruits were falling asleep, so after class, the whole platoon would head off to the sand pit and receive Incentive Training. Try to avoid this if possible.

Physical Fitness Training

In the beginning, you will be completely worn out, but it does not stop there. Usually when you are done with scheduled physical fitness training, the Senior Drill Instructor will take the platoon to the sand pit for Incentive Training.

Physical fitness training will consist of sprints, obstacle courses, ability runs, circuit courses, platoon runs, stationary runs and so on. PT is always changing, so there is growth and development throughout recruit training. Just when you think you have it figured out, they are going to flip it on you and shock your muscles even more. You will have the Initial Strength Test while in receiving, your first PFT around the end of first phase, and the final PFT towards the end of recruit training.

All the different training will help you to improve your overall run time. Remember, your score counts towards promotion. If you do not perform at maximum, you are only hindering your future promotion potential. Prepare yourself prior to arrival so it is not so hard. Use Boot Camp to improve your physical fitness, not to start it. Developing your physical fitness will increase your overall self-confidence as a recruit and future Marine.

The following chart helps score your PFT results.

Exercise	Points	Max Exercise	Max Points
Pull-ups	5 per pull-up	20 pull-ups	100
Sit-ups	1 per sit-up	100	100
Running	3 miles in 18 minutes Lose 1 point for every 10 seconds past 18 minutes		100

Example: A Recruit has 20 pull-ups, 100 sit-ups and a run time of 18:20. The PFT score will be 298 out of a maximum of 300.

PFT CLASS	Age 17 – 26	Age 27 – 39	Age 40 – 45	Age 46+
First Class	225-300	200-300	175-300	150-300
Second Class	175-224	150-199	125-174	100-149
Third Class	135-174	110-149	88-124	65-99

Marine Corps Martial Arts Program (MCMAP)

The martial arts program starts with some basic classes and then moves into lower body strikes, upper body strikes, falls and leg sweeps, knife techniques, armed manipulation, unarmed manipulation and counter chokeholds (to name a few). Each time you go out to MCMAP training, you will be in boots and utilities. You will run from the barracks to the training area where you will stage your war belts and water bottles. From there, you will be turned over to the MCMAP instructor, who is usually a black belt instructor and is also a Drill Instructor.

There are five different color belt levels with six degrees of black belts. You have tan belts, gray belts, green belts, brown belts and black belts.

As a recruit, you will be qualifying in the tan belt category. If you do not pass the MCMAP testing, you will not be able to continue training. These skills will help develop a warrior spirit on the battlefield.

Confidence Course

Just as it sounds, it builds confidence. It is a combination of approximately 11 unique challenges, which will challenge your strength, balance and determination in the development of self-confidence as a recruit. These will physically and mentally push each recruit. After some practice, recruits will negotiate the obstacles. Some of the obstacles are Arm Stretcher, Parallel Bars, Over and Under Logs, Skyscraper, A-Frame, and Slide for Life.

Confidence Course

You will conduct this course in boots and utilities after instructors explain the dynamics. Once the course is explained, you will be broken down into squads. From that point forward, you will start double timing (running) from one obstacle to the next. When you are waiting, you will be continuously double timing in place. Drill Instructors might take you over to conduct rope climbing while waiting on everyone else to complete the course.

Pugil Sticks Fighting

What are pugil sticks? Basically, they are huge Q-Tips that you use to fight the enemy (fellow recruits). You will be in boots and utilities for this event. When you get there, instructors will take everyone through the combat movement course. It consists of running with a rifle, stabbing hanging dummies with the weapon, jumping over walls, jumping into bunkers, low crawling and so forth. On my first cycle as a Drill Instructor, we had to demonstrate the course and I broke my right thumb when I jammed the weapon into a metal plate in the bunker. Make sure you do not overdo the event and wind up injured like I did because you will be dropped from your platoon.

Once you get through the course, you will start making your way over to the pugil stick fighting area. There, you will get dressed up with a flak jacket, groin protector, helmet and mouth piece. When your time comes, you will step into the ring against another recruit within 10 pounds of your weight. You will go into the ring and start swinging your pugil stick at the other recruit. When deadly blows land on your opponent, you will score a point. You will continue this for three major blows to determine who wins the battle.

This event is very important to Drill Instructors because Drill Instructors are competing against other platoons to see who has the toughest in the series. Some of you have not fought a day in your life. That is okay. Just go out swinging and get some. You will be nervous but it comes and goes just like anything else. If you get busted up, go down swinging! You will conduct pugil sticks I, II and III during recruit training in order to build your combat techniques.

SDI Inspection

The SDI Inspection is probably the most stressful inspection during recruit training. The platoon will form up in uniform with the service rifle. The primary purpose of this event is to get you used to conducting an inspection and being asked questions about what you have learned so far in recruit training. Drill Instructors from other platoons will come to your platoon, step in front, and have you do the following:

First, you will conduct inspection arms to ensure that your weapon is clear and free of any rounds in the chamber. Once you get it cleared, the Drill Instructor will snatch the weapon out of your hand and you will immediately drop your hands back down to the position of attention.

Second, you say the following "Good Morning/Afternoon/Evening Sir/Ma'am, Recruit __________ from ______________________, MOS_____ job __________." Then you will say either "Regular or Reserves Sir/Ma'am." A lot of kids mess this up and guess what? You are going to get yelled at for not knowing it. Drill Instructors are going to yell at you regardless of how poorly or how well you do it.

Third, the Drill Instructor will start asking you questions like your 1st General Order, Chain of Command, and Marine Corps History. They are trying to see if you are retaining what you are being taught in recruit training. They are also checking your military discipline, so regardless of how much you have or have not learned, maintain your military bearing.

Fourth, once the Drill Instructor has finished chewing you a new one, he will hand you back your weapon. Remember, there can only be two hands on the weapon at one time so you will grab it with one hand, not both. Once the DI removes his hand from the weapon, you will place your second hand on the weapon. Then you will send the bolt home, close the ejection port cover and take the weapon back down to the deck by conducting order arms. You will be taught these steps in greater detail in Boot Camp, but for now, this is good information to know.

Fifth, when he walks away, he is moving on to the next recruit to do exactly the same thing he did to you, put them under stress and see if they can remain calm under this intense pressure. Your heart will be pounding and you will be thankful he is gone. If at all possible, remain calm and confident. Everything is building you up to become a Marine. Learn from it and make sure you prepare for the next inspection better than you did for the first one.

Company Commanders Inspection

This inspection is similar to the SDI Inspection, with the Drill Instructors from other platoons coming through. The inspection might be in a different uniform depending on the location and season of the year. The Series Commanders and Company Commanders will be coming through platoons, inspecting the development of each platoon, and their overall performance and confidence. They will ask questions to see where you stand among the platoon. Here is a tip, if you pay attention to the inspector coming down your squad; they usually ask the same question, so make sure you are attentive. Those that fail this inspection will have to be re-inspected. You do not want that outcome. They will be inspecting your uniform, hygiene, knowledge, bearing, weapon and so on. Any RUST on your weapon is an automatic failure, so be sure you are not that person. If you fail, you are going to have a bad day. These two inspections are building you up for the last and final graduation requirement, which is the Battalion Commanders Inspection.

Battalion Commanders Inspection

This will be the final official inspection of recruit training. The other two inspections are building you up for this moment. If you do not pass, you do not graduate. This inspection is usually in Service "A" uniform. There will be a lot less yelling because you are close to graduation. The Drill Instructors have done everything in their ability to cause you to stress out, study, get ready for and perform on this last inspection, so it is really up to you.

Make sure you have the recruit to your left and right inspect you because recruits always seem to miss something. There will be time prior to uniform inspection to get yourself, your uniform and your weapon in order. Make sure you are wisely using the time that the Drill Instructors will give you to prepare for these inspections. Often, recruits fail because they just did not use their preparation time efficiently.

It is a little tough doing inspection arms in Service "A", so make sure you practice in uniform prior to the real event. Once again, any rust on the weapon is an automatic failure. You will notice in recruit training, when you start cleaning weapons outside, you are usually preparing for an upcoming inspection. This helps you see anything wrong with your weapon. Be proficient with your time. At the end of the day, Drill Instructors will do everything possible to prepare you for this inspection, so make sure you do your part in order to graduate.

Drill, Drill, Drill, Drill

This is the bread and butter of recruit training. From the first day you arrive until the day you graduate, you are drilling with or without weapons. There are so many drill movements that need to be taught with not enough time to teach them to perfection. Both recruits and Drill Instructors will be officially evaluated twice during Boot Camp. The first evaluation will be Initial Drill, which is conducted several weeks into training and is led by one of your Drill Instructors. The second evaluation will be Final Drill, which occurs towards the end of recruit training. Your Senior Drill Instructor will take your platoon on the parade deck.

Initial Drill

This will be the first time recruits are evaluated along with a Drill Instructor to see the progress of the platoon's instant, willingness, obedience to orders given during the conduct of drill. You will have a drill card with a halted portion, where you stand in place and do movements with the rifle, and a marching portion.

In order to prepare for Initial Drill, you will be practicing all day, every day. In the beginning, when you get your service rifle and you starting placing it on your right shoulder while marching, it is going to kill your bicep. It will hurt badly, but make sure you don't adjust because as soon as you do, the Drill Instructors will see you and Incentive Training will happen. You will drill to chow, class, PT and when you get back to the barracks in the evening. You will set up the squad bay to drill inside in what is called the "Mini Grinder". You will spend a lot of time and the recruits who are not performing will find themselves on the quarter deck or in the sand pit getting IT'ed.

Pay attention the best you can and try not to cause any undue attention on yourself. If you do not understand, sound off with confidence and let the Drill Instructors know. If not, fake it 'til you make it. Learning drill is not easy and there is not a lot of time for your team to get it right. When evaluation day comes, you have to be ready for any drill card the DI pulls. You will be evaluated by the Drill Master of your

Battalion and your company Chief Drill Instructors. Once again, if the platoon does not perform well, you are going to have a long couple of days and find yourselves in the sand pit quite often.

Drill

Final Drill

The Final Drill evaluation has been called the "Super Bowl". Drill Instructors pride themselves on drill because that is how we develop recruits to become one team. This event means everything to us. You could have been the worst platoon so far, but if you win final drill, you take the Super Bowl. The Senior Drill Instructor will be the one taking the platoon out for final drill. By this time in training, recruits are confident in drill and if they have been practicing, the platoon will show up and show off. There is nothing better than seeing a platoon in sync performing together as one when the commands are given. It is amazing! The last three platoons of mine won the Super Bowl. There is a lot of sweat and tears put into this, so make sure you motivate the other recruits to perform, however you can.

The following pages contain some example drill cards that platoons could conduct.

INITIAL DRILL CARD #1 EXAMPLE

1. FORM THE PLATOON
2. CLOSE/EXTEND ON LINE
3. OPEN RANKS
4. PARADE REST
5. PORT ARMS
6. LEFT SHOULDER ARMS
7. ORDER ARMS
8. RIGHT SHOULDER ARMS
9. PRESENT ARMS (FROM RIGHT SHOULDER ARMS)
10. SIDE STEP (LEFT/RIGHT)
11. CLOSE RANKS
12. BACK STEP
13. COLUMN RIGHT (HALTED)
14. CLOSE WHILE MARCHING
15. COLUMN LEFT
16. RIGHT FLANK RETURN TO COLUMN
17. LEFT OBLIQUE/ MARK TIME
18. COLUMN HALF LEFT (REPEAT)
19. CHANGE STEP
20. COLUMN LEFT
21. EYES, RIGHT
22. FORM FOR PHYSICAL DRILL
23. REFORM THE UNIT
24. DISMISS THE PLATOON
25. PLATOON APPEARANCE

INITIAL DRILL CARD #2 EXAMPLE

1. FORM THE PLATOON AT CLOSE INTERVAL
2. EXTEND ON LINE
3. OPEN RANKS
4. PARADE REST
5. PORT ARMS
6. RIGHT SHOULDER
7. RIFLE SALUTE
8. ORDER ARMS
9. LEFT SHOULDER ARMS
10. PRESENT ARMS (FROM LEFT SHOULDER ARMS)
11. ABOUT FACE (REPEAT)
12. CLOSE RANKS
13. CLOSE THE INTERVAL, COLUMN HALTED
14. EXTEND WHILE MARCHING
15. COLUMN RIGHT (HALTED)
16. RIGHT FLANK RETURN TO COLUMN
17. COLUMN RIGHT
18. LEFT OBLIQUE /HALT IN PLACE
19. COLUMN HALF RIGHT (REPEAT)
20. COLUMN LEFT
21. EYES RIGHT
22. FORM FOR PHYSICAL DRILL
23. REFORM THE UNIT
24. FALLOUT
25. PLATOON APPEARANCE

FINAL DRILL CARD #6 EXAMPLE

1. FORM THE PLATOON AT CLOSE INTERVAL
2. EXTEND ON LINE
3. OPEN RANKS
4. RIGHT SHOULDER ARMS
5. LEFT SHOULDER ARMS
6. PORT ARMS
7. ORDER ARMS
8. PARADE REST
9. PRESENT ARMS
10. SIDE STEP (LEFT/RIGHT)
11. ABOUT FACE (REPEAT)
12. COLUMN OF FILES (RIGHT)
13. COLUMN OF FOURS
14. COLUMN RIGHT (HALTED)
15. CLOSE WHILE MARCHING
16. COLUMN LEFT
17. LEFT FLANK RETURN TO COLUMN
18. RIGHT OBLIQUE / HALF STEP
19. COLUMN HALF LEFT (REPEAT)
20. COLUMN LEFT
21. MARCHING MANUAL
22. MARCH TO THE REAR (REPEAT)
23. EYES RIGHT
24. DISMISS THE PLATOON
25. PLATOON APPEARANCE

FINAL DRILL CARD #7 EXAMPLE

1. FORM THE PLATOON
2. STACK ARMS
3. TAKE ARMS
4. LEFT SHOULDER ARMS
5. RIGHT SHOULDER ARMS
6. RIFLE SALUTE
7. PORT ARMS
8. PRESENT ARMS (FROM PORT ARMS)
9. PARADE REST
10. SIDE STEP (LEFT/RIGHT)
11. CLOSE RANKS
12. ABOUT FACE (REPEAT)
13. COLUMN RIGHT HALTED
14. CLOSE WHILE MARCHING
15. LEFT FLANK/RETURN TO COLUMN
16. COLUMN LEFT
17. RIGHT FLANK/RETURN TO COLUMN
18. LEFT OBLIQUE/PLATOON HALT
19. COLUMN HALF LEFT (REPEAT)
20. COLUMN LEFT
21. CHANGE STEP
22. MARCHING MANUAL
23. EYES RIGHT
24. FALL OUT
25. PLATOON APPEARANCE

Swim Qualifications

If you do not know how to swim, I would strongly suggest that you take lessons prior to going to Boot Camp. This is a graduation requirement and you cannot move on unless you qualify. Swim Qual will build confidence in the water for each recruit. You are required to pass CWS-4. The basic techniques learned will be conducted in the utility uniform (cammies). Once you pass the basics, you will have a chance to advance to CWS-3, CWS-2 and CWS-1. The better you are able to perform, the more points you will get towards promotion, so make sure you keep going if you can. Once you get past CWS-4, you will move into full combat gear, rifle, helmet, flak jacket and pack so things do get more difficult as you go on. Prepare as much as you can to ensure you make it past the first round and meet the minimum requirements.

Marksmanship Training

By the time you get to marksmanship training, you have been in recruit training for quite some time. You have become familiar with the weapon and now are ready to roll into the fundamentals of marksmanship. This training will be conducted in two weeks, one week of grass week and one week of firing at known distances.

Grass Week

Grass week will be a series of classes given to the recruits by a Primary Marksmanship Instructor (PMI). These Marines are expert shooters and have developed a skill to put the bullet in the bullseye every time. The PMI will take you through step-by-step weapons training and teach you how to get the bullet in the bullseye. There are 4 shooting positions that you will be taught: standing, kneeling, sitting and prone. Some of these positions are going to hurt badly because you are not used to them.

You will spend a lot of time dry firing at the snap in circle. There will be barrels with different types of targets. You will learn how to make adjustments, change magazines, breathe and dry shoot properly. It will be a long week, but make sure you prepare because you cannot graduate without qualifying on the rifle range.

Firing Week

During firing week, some of you will be firing your weapons in the morning and the other recruits will be pulling the targets in the pit. Each recruit will be firing from 200, 300 and 500 meters. You will have several days to shoot the course prior to qualification day, which will be around Thursday or Friday. Every Marine is a basic rifleman and there are three shooting badges you can earn, which are marksman, sharpshooter and expert.

Field Training

This will be the beginning of recruit basic combat skills. Expect classes in field sanitation, camouflage, cover and concealment, basic warrior training, movement, hand and arm signals, land navigation, combat leadership, combat endurance course, map skills and setting up shelter half camps. During field training you will learn how to live in field conditions with many random events happening continuously. There will be hikes, classes and other evolutions that will challenge you.

Gas Chamber

This is an event that I believe makes everyone nervous. You see smoke coming out of the building that touches your skin and immediately starts making you sneeze and cough. Your eyes water and you are not even in yet. You see a group of recruits go in, the Senior Drill Instructor and officers are inside waiting for recruits to come through. You will follow the recruit in front of you into the building, where you will walk around until all the recruits are in. They will have you do several things inside the gas chamber to ensure your mask is sealed properly. Some examples will be side straddle hopes, bending over and shaking your head from side to side. Then, you will be asked to take a deep breath, close your eyes and remove your mask from your face.

Immediately you will feel the CS gas starting to burn any exposed skin, and when you remove your mask it will also burn your eyes, throat and nasal passageway. Although painful, it will convince you how effective your chemical suit and gas mask are because you did not feel these sensations until you removed your mask. It is also an indelible memory to take with you to combat of how deadly serious a real chemical or biological attack can be.

You want to make sure you DO NOT breathe because if you do, you are going to have a tough time. You have to remain calm. Some recruits start freaking out and try to run. Guess what, you are not getting out. You have to complete this evolution as a graduation requirement. Stay calm and do not breathe again. Be smart and calm because this is ultimately preparing you for a chemical or biological attack in combat.

Humping (Hiking)

We call hiking in the Marine Corps humping. You are humping whatever is on your back around with you. You will perform a forced march as a company with three, five and ten mile humps, with about 60 to 80 pounds on your back. The key to humping is keeping the formation tight by being able to touch the recruit in front of you and to walk in their footsteps. Also, when walking in loose gravel, make sure you try to walk flat footed so that you are not digging into the ground. This makes it easier to keep pushing. Recruits fall out left and right during humping but you can avoid this outcome. It truly depends on your personal determination and willingness. These humps will build you up over a couple of weeks to prepare for the Crucible, where you will hump nonstop over a three day period covering over 50 miles.

The Crucible

This is something I did not do as a recruit when I went through Boot Camp in 1995. Shortly after I graduated recruit training, the Marine Corps introduced the Crucible. When I was in the fleet, I thought the Crucible was going to be a joke. When I went through DI school to become a Drill Instructor, we had to do the Crucible. I can honestly tell you that it was extremely tough, especially if you put out and perform. The Crucible will be 54 hours long and will challenge you physically and mentally. Here are some facts about the Crucible:

- Recruits will travel around 50 miles.
- 36 stations with around 30 different problem-solving events.
- Only 3 MREs (Meals Ready to Eat) during the 3 day event.
- You will have your utilities, gear and service rifle with your day movement pack on your back.
- Day and night events will be conducted.
- Only about 3-4 hours of sleep a night.
- There will be challenging long humps, problem-solving events and team building stations in a rite of passage that will never be forgotten.

Team Week

This could be several different things depending if you are in San Diego or Parris Island. You could find yourself preparing for inspections, getting fitted for uniforms, practicing final drill or working around the base maintaining the appearance of the Depot. Whichever you find yourself doing, remember it is considered team week, meaning that you must work as a team to accomplish the goals at hand.

Graduation Requirements

These are provided below. In order to become a United States Marine, you must complete all requirements; otherwise you will be recycled and will not graduate with your platoon.

1. Pass all practical Marine Corps Common Skills Test with a proper grade.
2. Pass the Physical Fitness Test along with the height and weight standards.

3. Pass Combat Water Survival.

4. Qualify with the Service Rifle.

5. Pass the MCMAP Examination.

6. Successfully Complete the Crucible.

7. Pass the Battalion Commander's Inspection.

Graduation Week

You will find recruits very busy getting ready for graduation. There will be ceremony practices several times a day, uniform preparation and you will take care of any loose ends. By this time, recruits have been transformed from civilians to basically trained Marines that will soon be awarded the Eagle, Globe and Anchor.

Visitors Thursday

Every recruit will be excited about this moment and getting to see their loved ones. I remember during family day, it was so awesome to see everyone whom I had not seen since leaving for recruit training three months prior. The night before graduation, I had a chance to go through my items that were boxed up upon arrival at MCRD. It seemed like forever since I had last seen them. It felt that much better getting those items back. I earned it back!

Additional Notes

- If you cannot physically perform during recruit training, you will be placed in the Physical Conditioning Platoon (PCP). You will remain there until you are strong enough to continue training with a new platoon.

- If you get injured or fall sick during recruit training, you will be placed in the Medical Rehabilitation Platoon (MRP). There you will receive medical care and attention towards your recovery so that you can return to training.

- You will receive over 100 hours in Marksmanship Training.
- You will receive over 60 hours in Physical Training.
- You will receive over 50 hours of training during the Crucible.
- You will conduct over 40 hours of close order drill.
- You will conduct over 30 hours in MCMAP training.
- You will spend over 18 hours in field training.
- You will spend over 16 hours in Combat Water Survival training.
- You will spend over 15 hours in Core Values training.
- There are about 40,000 new Marines a year.
- If you display a negative attitude, you will pay for it.
- Lean on your fellow recruits when things get tough.
- Protect your feet during training. I left recruit training and my feet were completely numb because of all the standing in Boot Camp.
- Try and stay calm. A lot of people are going to be nervous. You can perform better if you stay calm.
- Remain professional regardless if you are happy with your Drill Instructors' performance. Remember, they are working their butts off to train the platoon.
- Try not to waste your money when you go to the PX. This money is yours and the less you spend, the more you have when you leave. Get what you need only.
- Do the right things for the right reasons.

7 AFTER BOOT CAMP

While Boot Camp and OCS are all about indoctrinating you into the military way of life, follow-on training is about teaching you a specific profession. You have to make it through Boot Camp just to have the chance to try out for your next school. There are far too many training pipelines in the military to provide an adequate outline of all of them. However, many of these schools are just as intent on washing out students, particularly in professions that involve the potential loss of expensive equipment (such as aviation) or heightened risk to life (such as Special Forces). As such, the pressure to perform at the highest level never really stops.

After Boot Camp, most recruits will get to go on leave before the next stage of training. For Marines, leave is for ten days before moving on to the Marine Combat Training (MCT) or the School of Infantry (SOI), and then on to Military Occupational School.

Boot Camp Leave

Most recruits return home with their families and loved ones and others go on a vacation. Regardless of where you go, everyone will notice a difference in you personally and professionally. Your friends will be the same but you will be different. People will notice the way you walk, the way you talk and the way you carry yourself is different from what they were used to before you joined the military. During my ten days of leave, I remember not really wanting to go back to the

Marines. However, I noticed that everyone was still the same. I had to go back. Plus, I definitely wanted an adventure, one that could not be found at home. Many of you may not want to return from leave, but you must. If not, everything for which you worked so hard will be for nothing.

Marine Combat Training / School of Infantry

For Marines, after your ten days leave, your next stop will be Marine Combat Training. This course is 29 days long and consists of training that fine tunes your basic level common skills needed during combat. Combat leadership, marksmanship, improvised explosive device training, defensive positions, combat formations, assaults, patrolling, humping, convoy formations and operations, patrols, and land navigation will be taught to every Marine.

If you choose an infantry military occupational specialty, you will be off to the School of Infantry, which is a 59 day course covering in depth training on how to fight and win in combat.

Military Occupational Training (Specialty Training)

Your experience in Boot Camp can sometimes make it easy to forget that you still will specialize in a particular vocation while serving in the military. In most cases, your specialty was chosen when you initially signed your contract with your recruiter. In some cases, your specialty can be changed based upon test results that certain recruits take during Boot Camp. Specialty training will be conducted by professionals who have years of military experience in their field.

Most military schools are less intense than Boot Camp (but are still intense nevertheless). You may find that these schools can be quite academically challenging and you will spend a lot of time studying. You are training to become experts at operating and managing equipment. As such, the expectation is that you know your equipment. These requirements never end. Even when on deployment, you may spend hours reviewing operating manuals just to refresh your memory.

Then there are some schools where there is no point in letting go of the Boot Camp mentality. If you go from Boot Camp to Special Forces or Navy Seal training, welcome to a new level of pain. In these cases, you will experience once-in-a-lifetime events, such as drown-proofing, endless days without sleep, physical training that will push you beyond your limits, and so forth. You will be yelled at just like in Boot Camp.

Off to specialty training.

The takeaway from any of these schools is that they should be considered a full-time obsession in order to succeed. If you had bad study skills in high school or college, it is not a good idea to import those habits into your military career. Boot Camp tries to break recruits from the civilian way of life, and you should carry-over the work ethic developed at Boot Camp to other areas of your military career (and life in general). If you treat your follow-on schools with the same intensity and ethic that you developed at Boot Camp, your chances of finishing those schools increase significantly.

Remember, the military can be ruthless in choosing which trainees pass and which ones fail. Getting accepted to and even making it through Boot Camp is just the beginning. A successful military career involves more than graduating Boot Camp. You also have to finish your remaining training and get completely qualified in your profession.

Only once you have qualified in a military profession do you have a chance to build a successful career, regardless of whether that career is 5 years or 20 years long.

Duty Station (Fleet)

Upon graduation from specialty school, you will receive orders to your first duty station. You may have the opportunity to request your choice of duty station. However, at the end of the day, you will be stationed at a location where the military needs you the most. Do not be surprised when you get a location that you did not request. My first duty station was Iwakuni, Japan, even though I had selected the United States East Coast. Once you get to your first duty station, surround yourself with positive colleagues. There will be a lot of negative and positive influences. The next three, four, or five years can be fun or it can be LONG. It really depends on you.

Some of the bad habits you had before Boot Camp will start to come back. You are not under the extreme control structure put in place at Boot Camp. You will have leaders who will control some aspects of your life, but you start to get liberty after working hours and on weekends when you are not in training or standing duty. Surround yourself with the people you want to be like. If you want to be successful, find some and hang out with them.

I look at it this way. I can apply myself, get paid more money and have less people telling me what to do. Or, I could not apply myself and watch others pass me, get paid less than I could, and be told what to do all the time. Because I was a focused Marine, I rarely found myself on details reserved for non-performers. Working parties happen all the time. I did not want to be part of those if at all possible.

Reporting to your first duty station can be an unusual experience. Prior to this point, the military has expended a lot of resources training you and in many ways you are the center of attention. However, the military expects a return on its investment, and your first real active duty station is where that payoff begins.

If your experiences through Boot Camp and follow-on schools have been surreal, your first command is where the military becomes all too real. You are now accountable for your actions in a real-world way. Your commanding officer and the other individuals in your immediate chain of command will expect you to take full responsibility for yourself, know your job and be able to work under enormous amounts of pressure.

If you are an officer, when you arrive at your duty station, you will be replacing an outgoing officer. If the transition is timed correctly, the departing officer will still be at the command when you arrive, and you will have a few weeks to shadow her. A smooth turnover period can make a big difference in your future success. You can use the opportunity to get a sense of your enlisted personnel, get tips on how the people above you work, obtain a layout of your equipment, and read through as many files and items of paperwork as you can handle. At some point you will be introduced to the people who will be working for you. The first introduction will set the tone for your tenure, as your enlisted personnel will start assessing you as a leader from the moment they meet you.

It is virtually impossible to provide any uniform guidance on how to adapt to your first command. The variety of scenarios you may or may not face depends on a nearly infinite number of factors. You might check-in to a department that is a complete disaster because the senior ranking enlisted member in charge has chosen to go on the "ROAD" program. "ROAD" means "Retired On Active Duty", and is military slang for someone nearing retirement who has decided to put in no more than the bare minimum effort for the remainder of his time.

Or you could arrive to a seasoned elite forces team where every person is a born leader and quintessential professional. Each circumstance will make demands on you. How you carry on in such a wide variety of situations cannot all be explained in this book. But regardless of the nature of your first and subsequent duty stations, you are now a cog in the military machine.

In the military, success means becoming proficient in at least all of the following:

Equipment. Become an expert in the operation and maintenance of your equipment.

Appearance. You are judged by how you look and carry on day-to-day. It is tempting to relax your standards after leaving Boot Camp, when there are no longer Drill Instructors around to yell at you for looking slovenly. But your colleagues will lose respect for you if your uniform is dirty, you are unclean, or you let your body turn soft.

Leadership. Never forget the importance of being a good leader. If your senior enlisted staff members are top notch professionals, learn from them and their leadership skills.

Paper Administrator. No matter how skillful you are as a leader and as an equipment operator, your success in the military just as importantly depends on your ability to manage paperwork. Keep your files organized and be able to reference documents efficiently. Remember, the military does not make the public aware of the amount of paper-pushing involved with serving, but you will spend as much time behind a computer as you will behind the scope of a rifle.

Regardless whether you are checking in to your first command or your tenth, anytime you are in a new situation, if you approach it with the Boot Camp mentality, you will build a foundation for future and lifelong success. If for the first three months of checking in to a new command, you dedicate yourself to the task of learning everything with the same mentality as during Boot Camp, your chances of success increases. Yes, this means working some 20 hour days, learning your job with passion, and focusing on nothing else but your mission.

Deployment to Combat Zone

Everyone in the military is eligible to deploy to areas of the world where active hostilities are ongoing. Combat duty can only really be described by the individuals who have experienced it first-hand. To give you a glimpse, this portion of the book recounts portions of an actual interview with a combat veteran.

Q. Describe your reaction when you first found out that you would be sent to a combat zone.

A. Apprehensive, I wasn't losing sleep over it, but a bit apprehensive, after 8 years of active duty and then a few years of reserve duty, to actually have to go to the show was interesting to say the least, you see the news every day, sailors dying, people dying, and you are going into the middle of all that, it makes you apprehensive.

You land in the middle of the night when you get there. They don't want you landing in the day. You have to wear a helmet, Kevlar vest, pistol, and sit in a C-117 until you land. When I arrived, it was pitch black. I was more afraid of not knowing where I was supposed to go and supposed to be rather than being scared of actually being there. A Captain I met there gave me a lift to my room and that was that. No sleep problems the first night. The next day, you realize you are in the middle of Camp Victory, you see out, you see the mosques, you see the buildings, the palaces, and the whole entire surroundings and you realize that you are in the middle of everything that has been going on for the last ten years.

And you wonder how the hell you got there.

Q. Describe the personal preparations you made prior to departure.

A. I moved some moneys around. I didn't need it. I wanted it in different places. My wife took care of most things. Looking back, it would have been nice to get a better grasp of the things that you are able to use now, such as the different programs available to those who go out there. Knowing more about that would be nice. You take it for granted. I'm not going to use these services now anyway, but you wish you knew beforehand, they give all the information to you on the back-end coming out, but all these programs are out there.

Q. Describe how your spouse, children, and/or family prepared for your departure.

A. My buddy was apprehensive to tell his wife. She was a little nervous at first but got used to it. My wife said, "It's no big deal to me". My mom, she was nervous, but once you get to that region of the world, and see the techniques we use, and the safety procedures that we implement … remember, the senior guys in the terrorist organizations don't want to die themselves, so they send these low-level terrorist knuckleheads out there to do their work for them. There are people who want to hurt us. But there are safety procedures out there. Everyone has a gun and they are all doing their time. Where I was in Iraq, except for the occasional crazy who snapped, you didn't have to worry. The average soldier out there now is not taking it to the enemy. Special Forces are operating in combat. Afghanistan is a different story because you have regular Marines taking it to the enemy and the Taliban.

Knowing this didn't necessarily put everyone at ease, but it made everyone take it for granted. Your family really doesn't know what it is. When you are there, you can describe it, but there will be apprehension with the family. People look at you differently when you say you spent a few months in Iraq. But the people who are there - did I ever feel like I was in peril? No. Was it interesting to drive in the city? Yes. Was it interesting to go the Prime Minister's building? Their version of Special Operations Command? Sure. Their generals? Yes. I never felt in danger *per se*. You take it for granted I guess.

Q. What happened that was unexpected?

A. End of May, beginning of June, incident with a soldier, that killed five other soldiers at the command counselor's office, and what took me by surprise is the fact that all the stories leaked about it on the base, whether the kid was suicidal and the command had taken him there, and the kid overpowered his chaperone and took his rifle, and killed another four people, and of course he was killed. He shot people randomly, and they took him into custody. My understanding is that he was suffering from Post-traumatic Stress Disorder, they took the weapon away from him, somewhere in there – they lost eyes on him, he went in there and started killing people, and they took him into custody. That incident kind of sucked.

The overall safety, I mean being there, you are within a few miles of the perimeter wall, yeah they can lob a mortar in on you, but they haven't done that in a long time – 3 or 4 months prior to my arrival was when the last mortar attack occurred, my buddy heard it.

When I drove through the city a couple to times, I knew it was going to be dirty. People there don't know how to adapt to change I guess. Some people sit around all day not trying, and they don't like others to succeed, so they blow stuff up.

Q. How did your mind react to continually being in a hostile environment?

A. There isn't a major difference between the palace and a vehicle. You are in a couple of million dollars' worth of machinery that can take a blast. The terrorists weren't aiming or targeting those things. The vehicles change their routes. A couple of years back, they fired on them. It is interesting driving throughout the city and seeing these cars next to you. Apprehensions, a little bit, but change in thought process, no.

Q. What advice would you give to someone going into a hostile area for the first time?

A. Take a good book. If you are on patrol, I would be a little more apprehensive. The terrorists are going out killing Iraqis. If you look at it, all the bombs you hear in the news target big marketplaces. If you are an American, when you go into the city, you will be in the city with a decent number of other people. Apprehension? Sure.

The Afghanis have been fighting since the creation of their country. All they do is fight. From the late 1970s when the Russians tried to take over, we are being shot at with the same stuff from 25 years ago.

Q. What did you face on hostile deployment that you would not otherwise face on regular deployment or on shore duty.

A. Beings stuck in one place was the hardest. That sucked.

Q. Please describe your typical daily routine?

A. My job had just been established within the last four months, and my buddy and this other officer were the first two. We would sit in the central operating area. Any information came from the forward operating area would come to us. If they needed help, we would send assets forward.

I came up with different war games, and say a foreign nation came over and shot our guys, what would we do as a response? We had to deal with rules of engagement, and determine which units stayed and which ones had to be pulled. It is enlightening to see how many people it takes to get something done.

8 DETAILS ON BECOMING AN OFFICER

Most individuals joining the military plan to enlist. But for some, becoming an officer is a career goal. This can occur either from the outset, or after serving a few years in the enlisted ranks. If you decide to become an officer, there are specific procedures and requirements that you need to follow, some of which are outlined here. Even if you are enlisting, you should read this chapter. You will be interacting with officers on a daily basis in the military, and knowledge of how they got there is beneficial to you.

Although there are several ways to become an officer, the primary three methods are the service academies, reserve officer training corps (ROTC), and officer candidate school (OCS). The services academies, West Point (Army), Annapolis (Navy and Marine Corps), the Air Force Academy, and the Coast Guard Academy are fulltime degree-granting colleges that also confer officer commissions to its graduates. ROTC units are attached to civilian colleges around the country with members working at their ROTC units once per week and throughout the summer. At the end of college, ROTC students are granted an officer commission.

Applicants to the service academies and ROTC are usually high school students and recent high school graduates. If you already have a college degree, you are unlikely to want to put yourself through another four years of school at an academy or ROTC unit just to get an officer commission.

If you already have a college degree, you become an officer through OCS. From the military's perspective, OCS serves a planning function. Since the academies and ROTC are four year programs, the military has to determine its manning requirements years in advance. Those numbers do not always work out. If there are manpower shortfalls in particular military specialties, OCS graduates can help fill those shortages in a cost effective manner (the academies and ROTC also involve four years' worth of costs to the military compared to three months' worth of costs for OCS graduates). The military also recognizes that some individuals who want to serve as officers have no desire to commit four years of their lives to the military prior to becoming officers. OCS provides an option to those individuals who want a quicker transition from the civilian world into the military.

With the OCS option, you can complete and graduate from college, and take the time to physically and mentally prepare for OCS. After you earn your officer commission, you go to follow-on schools in order to be designated with a particular military specialty. Eventually you report to your first command in order to take charge of your first division or platoon.

Military Branches and Specializations

Just like when enlisting, you have two major decisions to make when trying to become an officer. First, you have to decide which branch of the military you want to join: the Marine Corps, the Army, the Navy, the Air Force, or the Coast Guard. Then, you have to decide what job you want to perform once you receive your commission. You should make both decisions prior to approaching the military, as your choices will not only guide your early career path, but will also prepare you for dealing with the military recruiters who process your OCS application.

In deciding which branch and occupation to pursue in the military, take into consideration the following information.

The United States Navy

The Navy has revised its role in world affairs over the years, but the central premise of the Navy remains unchanged - to support the actions of its fleet of long-range ocean cruisers. The technology behind the fleet changes over time, and as such, impacts how the Navy conducts day-to-day business, but the mission of the Navy will always be command and control of the seas. The various specialties in the Navy generally cater to that mission. If you decide to join the Navy, your recruiter will offer you a particular specialty (referred to as a "designator") that will be your career focus as an officer. The major Navy designators are described below.

Surface Warfare Officers (SWOs) are responsible for the smooth operation of Navy ships. SWOs receive training in areas such as navigation, engineering, information systems, security, and deck operations. SWOs stand bridge watch when the ship is underway at sea ensuring safe movement from point to point. In most instances, the Navy turns to senior ranking Surface Warfare Officers when selecting commanding officers of ships.

Submariners perform the same general duties as SWOs, except submariners ply their trade in submarines operating at great depths in the ocean.

Pilots fly the Navy's various aircraft, including fighter jets, propeller driven planes, and helicopters. Many Navy pilots are expected to land airplanes on aircraft carriers.

Imagine yourself flying a fighter jet over the middle of the ocean. Blue skies and deep water all around. There is no land for hundreds of miles. You see a tiny dot bobbing in the waves. That is your lifeline. You have to land on that dot.

Naval Flight Officers (NFOs) provide airplane support to the pilots. NFOs are part of the flight crew and log as much time in the air as pilots, but NFOs are not tasked with flying airplanes. Instead, NFOs are responsible for much of the residual duties inherent in military aircraft, the types of duties that pilots cannot concentrate on when flying, such as manning radar or monitoring aircraft systems.

Navy Seals are the Navy's elite Special Forces. The Seals are involved in every sort of unconventional warfare imaginable, from diving, freefalling out of airplanes, jumping out of helicopters, land warfare, explosives, guerilla tactics, sniper action, and much more. Even during peacetime in the United States, the Seals see combat. The Seals are viewed as the most exclusive club in the Navy.

Explosive Ordnance Disposal (EOD) officers are specially trained Navy Divers who are responsible for both setting and rendering-safe explosives, both on land and in the water. EOD officers oftentimes work closely with the Secret Service to protect dignitaries, including the President.

Intelligence officers seek out enemy information that can be used to the Navy's advantage in both combat and day-to-day operations.

Supply Officers (SUPPOs) are logistical experts who ensure that all the supplies that the Navy requires to operate properly are in the right location, at the correct time, and in the proper quantities. SUPPOs are responsible for everything from ammunition to coffee.

Smaller Navy communities include Engineering Duty Officers, Public Affairs Officers, and Seabees (civil engineers responsible for rapid construction projects). There are other designators in the Navy that require advanced degrees, such as a law or medical degree. However, these professionals go through a different and more specialized commissioning route than the vast majority of officers.

The United States Marine Corps

The Marine Corps is an organization of epic, legendary proportions in the United States. It contains a potent mix of small size with a serious dose of elitism. Whether in its training, its recruiting, or in its day-to-day comportment, the Marine Corps exudes and demands intensity. The National Security Act of 1947 states that the Marine Corps is responsible for the seizure or defense of advanced naval bases and for the conduct of such land operations as may be essential for the prosecution of a naval campaign. But the Marine Corps, with its storied history interwoven into the fiber of this country, is so much more. While each branch of the military has elite forces, the Marine Corps in and of itself is oftentimes described as an elite force.

Marine Corps OCS instructors are on the lookout for candidates who really do not want to be Marines. The Marine Corps emphasizes being a Marine above anything else. The camaraderie that develops as a result of this focus is noticeable not only within the Marine Corps, but in the civilian world as well. When former Marines encounter each other, there is a kinship among them like no other.

But in order to join and succeed in this club, you have to be willing to pay the price. The Marine Corps lifestyle is demanding. The standard expected of every Marine is uncompromising, and this expectation shows not only in how each individual Marine carries on in uniform, but filters into off duty time as well. If you are the type of person that likes to let your hair down, the Marine Corps may not be your ideal choice. Once you are a Marine, you are always a Marine.

As an officer in the Marine Corps, you will be able to choose from many career options such as pilot or infantry. But if you are going to join the Marine Corps, do so because you primarily want to be a Marine, and as an officer, because you want to lead other Marines.

The United States Coast Guard

The Coast Guard is one of the most unique branches of the armed forces. During peacetime operations, the Coast Guard generally does not operate under the auspices of the Department of Defense, but instead works for the Department of Homeland Security. During wartime, the government has the option to transfer the Coast Guard to the Department of the Navy. Regardless, the Coast Guard is the military. Coast Guard personnel train at both Coast Guard and Navy schools on a regular basis. The Coast Guard fleet tends to operate and deploy closer to American territory than the Navy. It also is involved with drug interdiction and search and rescue operations off the United States coast.

Due to the smaller size of the Coast Guard, officers are expected to be more versatile. The Coast Guard does not designate its officers with the same particularity as the other branches. Instead, the Coast Guard divides its officers into specialties (and subspecialties).

Aviation	Safety, Operations
C4IT	Information Systems Management, Electrical and Electronics Engineering, Communications Management
Engineering	Ocean, Naval, Civil, Industrial, Aeronautical
Intelligence, Operations Afloat	Aids to Navigation, Ice Operations
Operations Ashore-Prevention	Vessel Inspections, Marine Investigations, Waterways Operations and Management, Port and Facility Safety and Security
Operations Ashore-Response	Defense Operations/Readiness, Boat Forces Operations, Maritime Law Enforcement, Ports, Waterways and Coastal Security Operations, Search and Rescue Coordination, Marine Environmental Response, Incident Management and Preparedness

The United States Air Force

The Air Force is the primary air and space military presence for the United States. The youngest of all the branches, the Air Force is by far the most technologically integrated department in the armed forces. If your aptitude is for electronic toys, the Air Force is certainly a viable career option. The Air Force provides long-range air capabilities to the United States. While Navy aircraft are usually launched from an aircraft carrier in the vicinity of a target, the Air Force is capable of launching aircraft from a base halfway around the world, refueling midair when necessary, conducting its operations, and returning to base. The Air Force also maintains a strong satellite presence in space, much of which assists in not only gathering intelligence from foreign installations, but to aid the Air Force's fleet of aircraft in conducting its long range missions. Officer career options (referred to as an "Air Force Specialty Code" or "AFSC") are broken down into numerous groups and subgroups based on a four-digit alpha-numeric identifying code. Major groups include pilot, navigator, space missile command control, intelligence, civil engineer, medicine and law.

Pilots fly both fighter jets and longer range aircraft. Air Force pilots are not required to land on aircraft carriers, but can be involved in missions that traverse half the world.

Navigators are similar to Naval Flight Officers. Air Force Navigators are responsible for non-pilot duties on aircraft.

The Air Force's Space Missile Command Control is responsible for United States military assets in space. These responsibilities include deploying satellites using various launch vehicles.

Another exclusive club is Combat Control, the Air Force's version of Special Forces. Combat Control officers are trained to perform all the exciting missions, including jumping out of airplanes, and diving. These specially trained warriors are also qualified FAA air traffic controllers. Combat Control deploys into hostile territory, secures the landing field, and then guides aircraft into the zone. Combat Control does not get as much media attention as other Special Forces, but their training and lifestyle is similar.

The Air Force also has fields such as Intelligence, Civil Engineer, Medicine, and Law.

The United States Army

The Army's role is land combat. Unlike the Marine Corps (which focuses its land operations from a leanly staffed amphibious assault perspective), the Army is a massive military machine that is designed not only to take a target, but to hold the target. Whenever the United States occupies a foreign territory, the primary occupation force is the Army. Due to its enormous size and massive scope of operations, there exists numerous occupations (referred to in the Army as an "area of concentration" or "AOC") from which an officer can choose.

Infantry has historically formed the backbone of the military. Infantrymen are soldiers whose primary mission is to conduct combat on foot. As an Infantry officer, you would be leading a platoon of soldiers who are trained in ground combat and small arms. Infantry is a physically arduous profession that involves long days and nights out in the field with very few modern day comforts. An Infantry officer's job is to lead troops in taking and holding a target. Army Rangers are an infantry special operations force.

Infantry forces are exposed to every personal toy and gadget imaginable, whether it is a pistol, sniper rifle, machine gun, or grenades and mortar. Today's Infantry are also equipped with high tech tools such as night vision goggles and precision GPS. The military is constantly developing ways to improve the Infantry with technology and you must become proficient in operating these tools.

Combat Engineers support Infantry by constructing (at times while under enemy fire) the various fortifications and other large infrastructure, such as bridges and roads, necessary to prosecute a successful ground campaign. Combat Engineers also clear minefields.

If you choose to become an Armor Officer, you will be in charge of units that drive tanks and other heavy ground fighting equipment.

The Army has both Field Artillery and Air Defense Artillery Officers. Artillery are the large weapons, such as rocket launchers, that cannot be carried by a single individual. As such, artillery is towed to its destination by various types of military vehicles. Field Artillery is tasked with stopping the forward movement of an enemy's large equipment, such as tanks, while Air Defense Artillery involves surface-to-air missile action designed to bring down enemy aircraft.

Like the Navy Seals, Marine Corps Force Recon, and Air Force Combat Control, Army Special Forces are trained in all aspects of unconventional warfare. Army Special Forces are more commonly known as Green Berets and deploy around the world performing covert activities behind enemy lines. Qualifying as a Green Beret can take up to two years and involves training in all facets of combat. Like their counterparts in the other branches, the Green Berets are very exclusive.

The Army has many other exciting specialties to consider as well, including General Aviation, Information Operations, Intelligence, Military Police, and Explosive Ordnance Disposal.

Joint Operations and Shopping Around

Many of the functions of the different armed forces branches overlap. The military has been consistently moving towards joint operations, where the services not only work together, but oftentimes train together. For example, Air Force Pilots train with Navy Pilots. This means that if you want to become a pilot and the Navy or Marine Corps cannot offer you a slot, you can approach the Air Force to check their availability. Or if you cannot be selected for Navy Explosive Ordnance Disposal, you can approach the Army. Of course, in order to have the flexibility to shop your services around, it is important to make your requests before you join the military.

9 SUCCESS IN THE MILITARY AND IN LIFE

Navigating the intricacies of a military career requires good judgment and planning. If you intend on making the military a lifelong profession, there are many accomplished senior enlisted and officers in the service who can provide mentorship.

If a long-term military career is not in the cards for you, it is imperative to achieve certain milestones while you are in the service. This helps you convert your military experience into potentially lucrative civilian opportunities. There are millions of veterans in the United States who have an institutional knowledge of the military and importantly, *who know which military careers can be defined as successful.* If your military experience is missing particular elements, most veterans will know that your career has question marks.

Your goal beyond Boot Camp is to have a military career that earns the respect of fellow veterans. That means there are a few objectives you must accomplish. In today's military, if you do not plan on completing the following four items, you may want to reconsider your decision to join.

- Finish Boot Camp (or OCS).
- Get qualified in a military specialty.
- Obtain the rank of E4 (enlisted) or O3 (officer).
- Deploy with an operational military unit.

Here are two examples for you to consider.

Example 1. You join the Marine Corps and five years from now, you are back in the civilian world interviewing for a job. The interviewer served in the Army. The following is a smooth interview.

Q. How long were you in the Corps?

A. Five years.

Q. That's awesome. Parris Island or San Diego?

A. OOH RAH! Parris Island.

Q What was your MOS? I myself flew helicopters for the Army.

A. Motor Transport. Best time of my life.

Q. Why did you get out?

A. Didn't want to stay in for 20 and five years worked for me.

Example 2. You join the Navy as an officer and six years from now are back in the civilian world interviewing for a job. The interviewer is an Air Force veteran asking questions about your time in the Navy. A smooth interview might go as follows.

Q. I see you were in the Navy. How long did you serve?

A. Five years.

Q. That's awesome. Were you ROTC or the Academy?

A. Actually, I commissioned through OCS.

Q. Oh, I heard that Navy OCS has Marine Corps Drill Instructors.

A It sure does, and let me tell you - those DIs are insane.

Q. I know. Those guys are intimidating. Anyway, I went through the Air Force Academy in Colorado, so the environment was a little different than Navy OCS. Tell me, what was your specialty? I was Intel, got out as a Major.

A. Supply Officer. I thought about staying in to see if I would get picked up for Lieutenant Commander, but you know when I left the Navy, I had been stationed in Bahrain for two years, it felt like the right time to come home.

Q. I can relate to that feeling. I thought about staying in for 20, but I kind of wanted to settle down and another major deployment was not in the cards for me. But let me tell you, I enjoyed the Air Force.

A I feel pretty much the same about the Navy.

Veterans love to reminisce about their days in the military. If any part of your story is incomplete or if you have to spend time trying to "explain away" missing elements of your career, most veterans will catch on pretty quickly that something is amiss with your record. To avoid any future uncomfortable questions about your military years, make sure you finish Boot Camp (or OCS), qualify for a specialty, get promoted to E4 or O3, and go on deployment.

Financial Planning

Relatively speaking, the United States military is not a bad paying job, particularly if you are in your late teens or early twenties. But the money can also be a double-edged sword. The steady income along with numerous benefits (including free health care, affordable life insurance, and tax-sheltered housing allowances) create a strong disincentive to save and budget.

Many new service members immediately purchase a brand new vehicle and saddle themselves with a car payment for five years. Because military income is steady, the risk of default on that car loan is minimal. However, many individuals become trapped in a vicious cycle where they become dependent on the military income.

If you definitively know that you plan to serve 20 years in the military and then collect retirement, you can afford to be a little lax with your budget in the early stages of your career. Where problems manifest themselves is with those who want to leave the military after four or five years to pursue civilian opportunities. Many individuals find that they *have to* stay in the military because they cannot risk losing the steady income, as they never saved any money.

Money is something that even the military neglects. Money is one of the most important factors to consider when making career plans. If you do not intend to serve for 20 years, it is important to remember the many benefits that disappear the moment you leave the military. For example, do you need to go to the doctor? You have the option to visit base medical. Does your family need health care protection? Sign up with TRICARE. Do you need groceries, gas or supplies? Head down to the Exchange and get goods for below-market prices. Retirement? Stay in the military for 20 years and receive a pension for life. Life insurance? Sign up for Service Group Life Insurance and have the premiums deducted from your monthly pay. Bills? If you live on a military base, power, water, gas, internet, and telephone are all provided. Moving expenses? One phone call and the military will take care of all your moving concerns.

These benefits disappear in the civilian world. Many service members confront money issues for the first time when they leave the military. Do you need to go to the doctor? Private health insurance becomes necessary for you and your family. The same goes for life insurance. When you need groceries and other supplies, the PX's subsidized prices are gone. Need cash? The best source of emergency cash will be your own savings account because no military organizations will be available to assist. Planning for retirement? Putting money away into a 401K or an IRA takes on new importance. If you need to move because you are changing jobs, you have to negotiate with movers on your own.

The moment you graduate from Boot Camp or OCS, it is highly recommended that you adopt a financial plan of action. Here is what to do.

Build a Reserve Fund

Begin making a concerted effort to save money. Building a nest egg of liquid cash should be your top priority by saving as much money as possible. There is no need to abruptly increase expenditures when you graduate from Boot Camp. Your savings account can grow significantly over a five year period. If you are busy putting maximum effort into your military career, living frugally should be second nature. Have the discipline to put money in your savings account, leave it there, and forget that it exists. Remember, establishing a reserve fund means putting your money in a savings account and does not mean putting your money into a mutual fund, CDs, stocks, or other creative investments, nor does it mean loaning money to anybody.

When planning a financial strategy, arguably nothing is more important than cash flow. There are corporations that exist for years that have never made a dime in profit but can still operate due to positive cash flow. Other companies have been forced into bankruptcy after running out of money despite the potential for future profits. The reason is that successful companies maintain positive cash flow (either by tapping into savings or accessing other outside sources of financing) to pay their bills and keep their operations afloat.

From your personal perspective, it is important to keep money in a liquid savings account for those periods of time when your income stream cannot pay your bills. In such an event, you can maintain a positive cash flow by using your savings. Consider the following scenario. You have $100,000 outstanding on your mortgage and $25,000 in your savings account. Your minimum monthly mortgage payment is $800. You could take the $25,000 and pay down your mortgage, because the interest on $75,000 is less than the interest on

$100,000, and your minimum monthly payment would be reduced to $750. But you could also keep the $25,000 in the bank.

Strongly consider keeping the $25,000 in the bank.

Suppose you took the $25,000 and partially paid off your mortgage. Now you have $0 in the bank and $75,000 outstanding on the mortgage. But what happens if you lose your job or are otherwise unable make the next scheduled payment? The credit bureaus are not going to remember that you prepaid $25,000. They will only remember that you missed a $750 payment. Eventually, the banks will foreclose on your house if you have enough missed payments. Now you are scrambling and under significant stress to avoid foreclosure and ruining your credit rating.

If on the other hand you kept the $25,000 in the bank and you lost your job, that next scheduled mortgage payment is easy to make. In fact, at $800 per month, you can make 31 monthly payments from the $25,000 in the bank. 31 months should be more than enough time to find a new job, stay stress free, and keep your financials in order. Importantly, the credit bureaus and the mortgage lenders will notice nothing because you are able to continue making your monthly payments.

This example typifies the importance of keeping a large reserve fund in an ordinary savings account. The reserve fund allows you to maintain a positive cash flow and pay your bills when your primary sources of income are taken away. The reserve fund also allows you to make a clearer decision with regards to your military career, since leaving the military takes away a major source of income. Having no money in a savings account makes it much more difficult to decide on career options than if you know you have one or two years' worth of living expenses put away.

Budget your Expected Day-to-Day Civilian Expenses

When you graduate from Boot Camp, envision for a moment that you were still a civilian. Take some time to forecast the expenditures you expect to incur as a civilian. Consider the costs outlined on the following chart for a useful guideline to planning monthly expenses.

Home	▪ Rent or mortgage and property taxes ▪ Water / Power / Gas ▪ Cable and Internet ▪ Telephone ▪ Home maintenance & improvement ▪ Home/rental insurance
Car	▪ Car payments ▪ Car insurance & registration ▪ Car maintenance & repair ▪ Gas
Personal	• Cell phone • Laundry • Food • Clothing • Entertainment • Travel • Life and Health insurance • Medical expenses and medicine costs • Saving for retirement • Personal care • Education costs • Children • Other, unexpected Costs

Plan Long Term Finances

Finally, envision beyond day-to-day regular expenses and assess your long-term goals. What other objectives do you want to achieve in life? Medical school or law school? Do you want to start your own business? When you contemplate these long-term objectives, first think outside of any monetary considerations and then decide and plan how to finance these endeavors.

By developing a financial strategy revolving around building a reserve fund, knowing your exact monthly expenditures, and planning long term finances, you can turn five years of military service into a springboard for virtually unlimited future potential.

A Few Words About Indebtedness

Never borrow money to fund your day-to-day living expenses. These expenditures should only be funded through your income (and through your savings during those periods when you have no income). Borrowing money should only be considered for long-term ambitions, such as going to school (for certifications in high demand). Even in these situations, if you can avoid borrowing money, always choose that option.

Borrowing money entails risks that cannot always be foreseen. For example, even if you purchase a house by using a mortgage – generally considered a prudent move – you are stuck with that mortgage until you (1) pay off the mortgage, (2) line up a buyer for the house, or (3) default and go into foreclosure. What happens if you need to move because of your job? The mortgage now becomes a weight. If you want to avoid ruining your credit rating, you have to keep your lenders happy. In order to pursue your career plans, you may be forced to make mortgage payments on an empty house if you are unable to line up a buyer.

Paying back debt takes longer than anticipated. A lot of financial institutions give the impression that repaying debt is not that difficult. For example, if you have a mortgage for $100,000 and work in a job that pays $120,000, could you theoretically pay off the mortgage in one year? Probably not. That job paying you $120,000 is actually paying much less after taking into consideration deductions for federal and state tax and social security. When you also deduct living expenses from your salary and add accrued interest to your loan, it would actually take many more years before you can pay off the amount you borrow. You might find out after a few months that the need to pay off your debts governs most of your decisions. That is a surefire recipe for a life of unhappiness.

Careful With (So-Called) Expert Advice

Many service members can relate a story of receiving a cold call from an investment "expert" trying to sell mutual funds or other creative financial instruments. Remember, it is no secret that those in the military have a steady (and highly disposable) income. There are more than enough shysters, con artists and thieves who are happy to help military service members part with their money. It will usually start with a telephone call at an inconvenient time from a friendly voice inviting you to lunch or dinner in order to explain to you the importance of saving for retirement. Once you are a captive audience, you quickly discover that "saving for retirement" involves investing your money with the person who just bought you lunch or dinner. These people are so good at their sales pitch that some of the sharpest people are taken for a ride.

The financial guidance provided in this book, especially the idea of keeping a substantial amount of money in a low-interest savings account, may be contrary to advice given by financial experts. But remember, the financial industry and all of its advisors, stock brokers, and mutual and hedge fund managers partly rely on an air of mysticism.

But sometimes even money making "superstars" turn out to be nothing but con artists. If you want to stay focused on your profession rather than be pre-occupied with such things as daily swings in the stock market, refrain from putting your money in risky schemes, or in places where withdrawal-restrictions may keep you from accessing money when you need it. You will have plenty of time to invest what you earned in your lifetime. Now is not the time.

10 CLOSING THOUGHTS AND STUDY MATERIALS

By joining the military, you embark on a path that opens many options to you both personally and professionally. If Boot Camp or OCS is the start of a 20 year military career, you could very well become a senior enlisted person or a ranking flag officer, even advising major commands and the Joint Chiefs of Staff. Or you may decide that five years in the military is sufficient and move on to pursue other goals in the civilian world.

Either way, the decision to join the military is life changing. You will never be the same after this experience. For me, I would do it again in a heartbeat. I wish you all the success in your military career.

Remember, *Never Quit and Never Give Up*!

This page intentionally left blank

THE CHAIN OF COMMAND

Unit Executive Officer
Unit Commanding Officer
Squadron Command
Chief of Operations
Branch Secretary
Secretary of Defense
Vice President
President of the United States

GENERAL ORDERS

1. To take charge of this post and all government property in view.

2. To walk my post in a military manner, keeping always on the alert, and observing everything that takes place within sight or hearing.

3. To report all violations of orders I am instructed to enforce.

4. To repeat all calls from posts more distant from the guard house than my own.

5. To quit my post only when properly relieved.

6. To receive, obey and pass on to the sentry who relieves me, all orders from the Commanding Officer, Command Duty Officer, Officer of the Deck, and Officers and Petty Officers of the Watch only.

7. To talk to no one except in the line of duty.

8. To give the alarm in case of fire or disorder.

9. To call the Officer of the Deck in any case not covered by instructions.

10. To salute all officers and all colors and standards not cased.

11. To be especially watchful at night, and, during the time for challenging, to challenge all persons on or near my post and to allow no one to pass without proper authority.

CODE OF CONDUCT

1. I am an American, fighting in the forces which guard my country and our way of life. I am prepared to give my life in their defense.

2. I will never surrender of my own free will. If in command, I will never surrender the members of my command while they still have the means to resist.

3. If I am captured, I will continue to resist by all means available. I will make every effort to escape and aid others to escape. I will accept neither parole nor special favors from the enemy.

4. If I become a prisoner of war, I will keep faith with my fellow prisoners. I will give no information or take part in any action which might be harmful to my comrades. If I am senior, I will take command. If not, I will obey the lawful orders of those appointed over me, and will back them up in every way.

5. When questioned, should I become a prisoner of war, I am required to give only my name, rank, service number, and date of birth. I will evade answering further questions to the utmost of my ability. I will make no oral or written statements disloyal to my country and its allies or harmful to their cause.

6. I will never forget that I am an American, fighting for freedom, responsible for my actions, and dedicated to the principles which made my country free. I will trust in my God and in the United States of America.

ENLISTED RANK STRUCTURE

	Army	Navy / Coast Guard	Marine Corps	Air Force
E1	Private	Seaman Recruit	Private	Airman Basic
E2	Private	Seaman Apprentice	Private 1st Class	Airman
E3	Private 1st Class	Seaman	Lance Corporal	Airman 1st Class
E4	Corporal/ Specialist	Petty Officer 3rd Class	Corporal	Senior Airman
E5	Sergeant	Petty Officer 2nd Class	Sergeant	Staff Sergeant
E6	Staff Sergeant	Petty Officer 1st Class	Staff Sergeant	Technical Sergeant
E7	Sergeant First Class	Chief Petty Officer	Gunnery Sergeant	Master Sergeant/ First Sergeant
E8	Master Sergeant/ First Sergeant	Senior Chief Petty Officer	Master Sergeant/ First Sergeant	Senior Master Sergeant/ First Sergeant
E9	Sergeant Major/ Command Sergeant	Master Chief Petty Officer	Master Gunnery Sergeant/ Sergeant Major	Chief Master Sergeant/ First Sergeant
E9	Sergeant Major of the Army	Master Chief Petty Officer of the Navy/ Master Chief Petty Officer of the Coast Guard	Sergeant Major of the Marine Corps	Chief Master Sergeant of the Air Force

OFFICER RANK STRUCTURE

	Army	Navy / Coast Guard	Marine Corps	Air Force
W1	Warrant Officer 1	Warrant Officer 1	Warrant Officer 1	None
W2	Chief Warrant Officer 2	Chief Warrant Officer 2	Chief Warrant Officer 2	None
W3	Chief Warrant Officer 3	Chief Warrant Officer 3	Chief Warrant Officer 3	None
W4	Chief Warrant Officer 4	Chief Warrant Officer 4	Chief Warrant Officer 4	None
W5	Chief Warrant Officer 5	Chief Warrant Officer 5	Chief Warrant Officer 5	None
O1	Second Lieutenant	Ensign	Second Lieutenant	Second Lieutenant
O2	First Lieutenant	Lieutenant Junior Grade	First Lieutenant	First Lieutenant
O3	Captain	Lieutenant	Captain	Captain
O4	Major	Lieutenant Commander	Major	Major
O5	Lieutenant Colonel	Commander	Lieutenant Colonel	Lieutenant Colonel
O6	Colonel	Captain	Colonel	Colonel
O7	Brigadier General	Rear Admiral Lower Half	Brigadier General	Brigadier General
O8	Major General	Rear Admiral Upper Half	Major General	Major General
O9	Lieutenant General	Vice Admiral	Lieutenant General	Lieutenant General
O10	General	Admiral	General	General
*	General of the Army	Fleet Admiral	None	General of the Air Force

* These officers are only activated during periods of war.

BOOT CAMP TERMINOLOGY

Military Term	Translation
Adrift	Scattered About
Aye Aye	Acknowledgment of Order
Bulkhead	Wall
Carry On	Resume Former Activity
Chow	Food
Deck	Floor
Ears	Lister to Drill Instructor
Eyeballs	Look at Drill Instructor
Field Day	Barracks Cleanup
Go Fasters	Running Shoes
Hatch	Door
Head	Bathroom
Ink Stick	Pen
Ladderwell	Stairway
Lead Stick	Pencil
Moonbeam	Flashlight
Overhead	Ceiling
Passageway	Hallway
Port	Left
Porthole	Window
Rack	Bed
Rain Room	Shower
Starboard	Right
Swab	Mop

ABOUT THE AUTHOR: SSGT NICHOLS

SSGT Michael Nichols is the Senior Drill Instructor in *Black Friday: Dark Dawn*, the film series about United States Marine Corps Boot Camp. SSGT Nichols joined the United States Marine Corps through the Buddy Program and graduated from Boot Camp on May 5, 1995. He then attended Marine Combat Training at Camp Pendleton and subsequently, Motor Transportation School.

SSGT Nichols checked in to his first duty station as a Private in Iwakuni, Japan. He was first promoted to Private First Class and then meritoriously promoted to Lance Corporal. SSGT Nichols moved to Aviation Ground Support Element 29 Palms, California where he was promoted to Corporal and then Sergeant.

SSGT Nichols was then reassigned to 1st Battalion, 7th Marines (1/7) at 29 Palms. After serving with the infantry for two years, SSGT Nichols received orders to MCRD San Diego to attend Drill Instructor School. Upon graduating, he was assigned to 1st Battalion, Charlie Company where he trained eight platoons to become United States Marines. It was during his seventh platoon that SSGT Nichols' life would forever change when Platoon 1141 was filmed and shared with the world. Not knowing it at the time, the filming of Platoon 1141, featured in the subsequent release of *Black Friday: Dark Dawn*, would lead people from all around the world to get to know SSGT Nichols and the Drill Instructor team.

In 2004, SSGT Nichols became the Protocol Chief for the MARFORPAC 3 Star Commanding General. SSGT Nichols deployed to Iraq in 2006 in support of Operation Iraqi Freedom. Based in Al Asad Iraq, SSGT Nichols completed 27 combat missions, 13 as a Platoon Sergeant and 14 as the Platoon Commander after his Lieutenant was severely injured in an IED attack.

After successfully serving in the United States Marine Corps, SSGT Nichols returned to his home state of Louisiana to pursue a business partnership, owning and operating a chain of convenience stores for two years. SSGT Nichols then moved back to California to launch two companies, Moto Entertainment and MDI 8.

Since leaving the Marine Corps, in addition to owning ten convenience stores and founding two companies, SSGT Nichols has appeared on television and produced several feature length motion pictures. SSGT Nichols is actively involved in serving Veterans and serving his community as a CARE Pastor in Southern California.

SSGT Nichols holds a Bachelor of Science in Business Management.

INDEX

Made in the USA
Monee, IL
06 November 2026

68085972R00090